U0322918

Green Walls and Rooftop Gardens

Growing Architecture

立体景观

（新西兰）格拉哈姆·克利里 编　李 婵 译

辽宁科学技术出版社

图书在版编目（CIP）数据

立体景观 / （新西兰）克利里编；李婵译. -- 沈阳：辽宁科学技术出版社，2014.8
ISBN 978-7-5381-8631-4

Ⅰ. ①立… Ⅱ. ①克… ②李… Ⅲ. ①城市景观—景观设计 Ⅳ. ①TU-856

中国版本图书馆CIP数据核字(2014)第103096号

出版发行：辽宁科学技术出版社
（地址：沈阳市和平区十一纬路29号 邮编：110003）
印 刷 者：利丰雅高印刷（深圳）有限公司
经 销 者：各地新华书店
幅面尺寸：225mm×285mm
印　　张：16
插　　页：4
字　　数：50千字
印　　数：1～1500
出版时间：2014年 8 月第 1 版
印刷时间：2014年 8 月第 1 次印刷
责任编辑：殷文文 陈慈良
封面设计：曹 琳
版式设计：曹 琳
责任校对：周 文
书　　号：ISBN 978-7-5381-8631-4
定　　价：268.00元

联系电话：024-23284360
邮购热线：024-23284502
E-mail: lnkjc@126.com
http://www.lnkj.com.cn
本书网址：www.lnkj.cn/uri.sh/8631

Green Walls and Rooftop Gardens

Growing Architecture

立体景观

（新西兰）格拉哈姆•克利里 编　李 婵 译

辽宁科学技术出版社

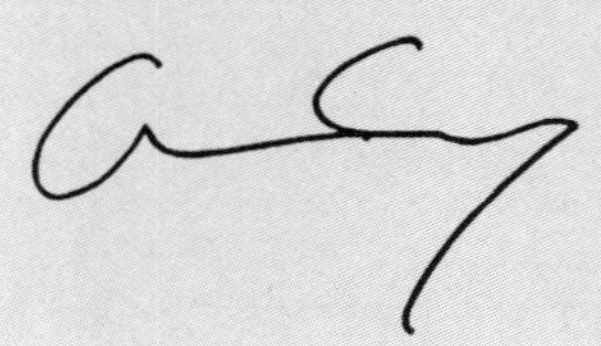

Graham Cleary

Graham Cleary founded Natural Habitats in 1978 in Auckland as a fledgling design build company under the mentorship of the city's then only Landscape Architect.

Graham has worked on and led many outstanding landscape projects in the South Pacific, including resorts, golf courses, commercial projects and exquisite private estates & homes. Under his leadership, Natural Habitats has grown to be the leading provider of integrated landscape management within Australasia. Graham demonstrates coherent leadership of multi-disciplinary teams in order to achieve the innovative, high-quality outcomes that Natural Habitats is renowned for.

Graham is also at the forefront of thinking on the use of natural systems to mitigate urban development issues. He played an essential role in Natural Habitats pioneering green roofs and wall technology within New Zealand.

Graham has served as consultant and advisor to numerous landscape design and environmental groups. He is a founding member of Landscaping New Zealand, an accredited Green Roof Professional, an executive of Green Roofs Australasia, an associate of KEA, part of the Great Barrier Trust, a member of Forest & Bird and of the Native Forest Restoration Trust.

Green Walls and Rooftop Gardens: Green Lungs Assisting in Sustaining Us

绿墙与屋顶花园——城市生活的源泉

Nature enriches our spirit, vitalising our souls and supporting our lives. Sadly, "Nature" is in retreat from our modern world due to urbanisation, farming and many other human activities. Without "Nature", we as a species, and many others will die.

The advent of modern green roofs and walls has allowed us to bring more of nature into even the most densely populated cities. Not only do these green pockets allow us to enjoy a slice of living Nature, they act as lungs assisting in sustaining us. They can become playgrounds in the sky for our children and ourselves!

No one can argue clean oxygen for us to breathe is a bad thing; it is a necessity. Nor could we argue safe green havens are not a massive benefit for our children in modern dense cites.

Green walls and roofs can also clean grey water and partially treated black water. This takes some of the load off cities stormwater and sewer infrastructures, improving downstream water quality and lessening the costs of required infrastructure.

Green roofs are morphing into "blue roofs" as people actively use this simple, well-tested technology. Green walls development has been a little slower in this area but the potential is just as large. Green walls can act as Vertical Wetlands with the same benefits, but at a fraction the ground foot prints required (saving money on valuable inner city space).

自然能够丰富我们的心灵，激活我们的精神，维系我们的生活。不幸的是，由于城市化和农业的发展以及许多人类活动，自然正在从我们的现代世界中退去。没有了自然，我们人类和许多其他生物一样，都将消逝。

现代绿色屋顶和绿墙的出现让我们能够在即便是人口最为密集的现代都市中注入更多的自然气息。这些绿化工程体量虽小，但是不仅让我们得以享受一点“活的自然”，而且是我们得以生存的基础、源泉。绿色屋顶可以成为孩子们的空中游乐场——对成人又何尝不是呢！

能够呼吸清洁的氧气对我们来说绝对是一件好事，这是无可争辩的，氧气是我们生存的必需品。在现代拥挤的城市中，安全的绿色游乐天堂对孩子们来说是多大的一笔财富，这也是不争的事实。

绿色屋顶和绿墙还能净化灰水（生活用水中污染较轻可以再次利用的水），并对废水起到部分的处理作用。这能够缓解城市排水系统的压力，改善下游的水质，并降低相关基础设施所需的费用。

随着相关技术的不断进步，绿色屋顶正在逐渐变成“蓝色屋顶”——屋顶水池的运用越来越普及了。在这方面，绿墙的发展相对来说较慢，但是开发潜力同样巨大。绿墙可以成为“垂直湿地”，跟运用水景的屋顶具有同样的益处，同时又不占用土地——对于寸土寸金的市区来说，这相当于带来巨大的效益。

格拉哈姆·克利里

格拉哈姆·克利里，1978年在奥克兰初创“自然栖息地”景观事务所。作为一家羽翼未丰的新公司，“自然栖息地”得到了当时奥克兰唯一一位景观建筑师的指导。

创办公司后，克利里主持设计了南太平洋地区多个杰出的景观项目，包括度假村、高尔夫球场等商业项目以及精致的私人房产和小型住宅。在他的领导下，“自然栖息地”景观事务所已经成长为大洋洲地区提供一站式景观设计服务的领军公司。“自然栖息地”一贯奉行跨学科的团队合作，以高品质的创新景观设计而闻名。

除了设计实践之外，克利里还站在理论和思想的前沿，倡导将自然生态系统引入城市环境，解决城市发展中的种种问题。“自然栖息地”事务所开发的创新绿色屋顶和绿墙技术走在新西兰前列，克利里在其中起到至关重要的作用。

克利里在多个景观环境设计团队中担任顾问和指导。他是新西兰景观协会的创建成员、公认的绿色屋顶专家、澳大拉西亚绿色屋顶协会执行主席、KEA公司合伙人、大堡礁基金会成员、新西兰森林与鸟类协会成员以及天然森林修复信托基金会成员。

Green walls can also be used as air filters, essentially being the front end of a building air conditioning plant. By drawing air through the wall, micro flora and fauna on the plants roots strip and consume volatile organic toxins, giving cleaner, healthier air quality.

Green walls and roofs are doing these amazing things more constantly and mostly for free! The benefits are multiple and multiply with more we establish around us.

Local habitats and eco systems are created and populated in the roofs and walls we form. These assist and support populations of birds, animals and insects often under immense habitat pressure from our own activities.

This book celebrates a sampling of some of the new works from companies like Natural Habitats who have been pioneering the development of beautiful green walls and amazing green roofs.

All of the projects are significant; some are pushing design boundaries and mixing art with function. This thankfully shows we are through the pioneering stages of green technology. Green roofs and green walls can now blossom in our cities and enrich us all.

I trust you enjoy and are inspired by the book.

绿墙还有过滤空气的作用，应用在建筑表皮上，是一部天然的空调。空气经过绿墙后，绿墙上的微型植物群落能够吸收挥发性的有毒有机物，产生更清洁、更健康的空气。

绿色屋顶和绿墙不断为我们做着如此巨大的贡献，并且几乎是免费的！它们带来多重益处，随着我们开发越来越多的绿色屋顶和绿墙，益处还将更多。

在我们建立的绿色屋顶和绿墙上，会逐渐形成生物栖息地和生态系统。那些在我们人类活动的干扰下面临巨大栖息压力的鸟类、动物和昆虫可以在那里安家。

本书选取了最新的绿色屋顶和绿墙项目，都是出自像“自然栖息地”景观事务所这样的知名公司之手的设计作品，代表了绿色屋顶和绿墙开发的最前沿。

书中所有的项目都有其独特价值，有些先锋的作品还开拓了绿化设计的边界，将功能与艺术完美结合。这些作品充分显示出我们正在绿色技术的开发之路上阔步前进。绿色屋顶和绿墙现在可以在我们的城市中绽放花朵，丰富我们的城市生活。

我相信读者定会享受本书的阅读过程，并从中收获灵感。

Sanitas P.

Sanitas Pradittasnee

Current Position
Director at Sanitas Studio

Education
1997-2001: Bachelor degree of Landscape Architecture, Faculty of Architecture, Chulalongkorn University, Thailand
2007: Master degree of Fine Art, Chelsea College of Art & Design, London, UK

Registration
Registered Landscape Architect, TALA No.190

Academics
2010-2011: Special Lecture on "An Application of Arts in Landscape Architecture", Department of Landscape Architecture, Chulalongkorn University, Bangkok, Thailand
2009 – present: Special Tutor in Master degree of Landscape Architecture, Department of Urban Planning, Silapakorn University, Bangkok, Thailand

To Live with Nature

与自然共生

The relationship between nature and human has been embedded for a long time. Human beings live in the realm of nature; they are constantly surrounded by it and interact with it. This presence of nature in an ideal, materialised, energy and information form in Man's Self is so organic that when these external natural bases disappear, Man himself disappears from life. If we lose nature's appearance, we lose our life.

In modern world, Man and nature interact dialectically in a way that, as civilisation develops, Man tends to become less dependent on nature directly, while his dependence grows indirectly. While Man is knowing more about nature and on this principle of transforming it, Man's power over nature continuously increases, but in the same process, Man comes into more and more considerable and profound contact with nature, bringing into the area of his activity growing quantities of matter, energy and information.

As society develops, the condition of living has changed and the unit of space has to be shared with more people. Every space is crucial for making a better living space and the efficient use of space. Though people are served by the convenience of city life, we subconsciously desire to live with nature, which we can see from the current trend of green design everywhere. By all modern technology and knowledge, we are equipped with new tools and method in dealing plantings and how to maintain them in different conditions. At the end, we have developed a new type of garden in a way it has never been used in the past.

This book is inspired by the image of acquiring nature in the modern world as an integration of planting and contemporary living condition and how to live with nature in the city. With high density of living in all urban area, the landscape design has been developed to be suitable with a new condition, yet still provide the notion of nature and its idea of paradise

人类与自然自古以来就是相辅相成、不可分割的关系。人类生活在自然界中，无时无刻不被自然所包围，并与自然相互作用。自然存在于人类理想的“自我”中，不论是精神上还是物质上，能源上还是知识上，我们都离不开自然。如果外部的自然基础消失了，那么人类自身的生命也就不存在了。如果我们失去了自然，我们也就失去了生命。

在现代世界中，人与自然辩证地发生相互作用，彼此相依。随着人类文明的进步，似乎表面上人类不再那么直接依赖于自然了，但是其实间接的依赖却更深了。人类对自然的了解越来越多，改造自然的本领也越来越大，凌驾于自然之上的能力不断增强，与此同时，人类与自然的联系也愈加紧密了，人类的活动领域涉及越来越多的知识与能源。

随着社会的发展，人们的生活条件发生了巨大变化，单位空间要与越来越多的人共享。打造更好的生活环境，提升空间的有效利用，每一个小空间都不容忽视。尽管城市生活带给我们许多便利，但是我们却下意识地渴望生活在自然中，这从当前随处可见的绿化设计趋势中可见一斑。现代化的技术和知识是我们的新工具，让我们能够应对不同条件下植物的栽种和维护问题。最终，前所未有的新型城市花园诞生了。

本书发端于“让自然回归现代世界”的美好愿景，探讨如何在城市环境中开展绿化，如何让植被美化我们的生活环境，以及如何让自然与城市共生。面对市区环境高密度生活的挑战，景观设计身负改善生存条件的重任，让“伊甸园”的概念重新回归城市。每个花园都是一座想象中的伊甸园，它

萨尼塔·普拉迪塔尼

目前职位
萨尼塔设计工作室经理。

教育背景
1997年-2001年：泰国曼谷朱拉隆功大学建筑系景观建筑专业学士学位；
2007年：英国伦敦切尔西艺术与设计学院美术专业硕士学位。

职业资历
注册景观建筑师（注册号：190）

学术背景
2010年-2011年：在朱拉隆功大学景观建筑系做“艺术在景观建筑中的应用”特别讲座；
2009年至今：在泰国艺术大学城市规划系任景观建筑专业硕士生特别导师。

garden. Every garden is an imaginary paradise; it reflects the character of location and the person who creates it. Through the history of garden from Mesopotamia to present day, the garden has been developed from wild field, agriculture park, royal park, allotments, landscape park and city park. In present day, the green architecture, green wall and roof garden become the main feature in contemporary urban landscape design including interior design. This illustrates how paradise garden can be integrated in urban landscape and the passion to live with nature.

As a landscape designer, our focus is on the balance between nature and human. “Not trying to create nature, as no one could, but framing it and bringing it closer to people.” In our practice, there is no clear boundary between building and garden, and we believe in three-dimensional green design to create a better experience in the city. In order to provide the green design, the green wall and roof garden have to be planned and designed together with the overall master plan from the beginning and allowing the space for the system to work properly. Not only the planting design, there is watering system and maintenance issue to make the green design be sustainable. In modern world, we can see more and more green walls and roof gardens everywhere. The green design has developed from being decoration wall to be part of architecture and overall master plan. Designing it, we need to consider it as part of the whole design system.

This book, *Growing Architecture: Green Walls and Rooftop Gardens*, developed from two main elements of three-dimensional green: vertical garden and roof garden. Apart from illustrating the current green designs around the world, the book provides the technical aspect of the green design, in term of plant selection, irrigation systems and maintenance. Designing it and sustaining it are the keys to make the growing architecture grow both aesthetically and functionally.

反映了每个特定地点的特色，也体现出创造它的设计师的个性。花园的历史可以从美索不达米亚追溯到现在，从旷野、农田、皇家公园、景观公园到现在的城市公园。如今，绿色建筑、绿墙和屋顶花园已经成为现代城市景观设计（包括室内设计）的一大特色。这充分显示了伊甸园是完全可以融入城市景观的，也反映了人类是多么渴望回归自然。

作为景观设计师，我们要将重点放在自然与人类的平衡上。“不是创造自然——没有人能创造自然——而是让自然回归人类视野，拉近人与自然的距离。”在我们的设计中，建筑与景观之间并没有清晰的界限，我们相信通过立体的景观设计能够带来更好的城市空间体验。为了实现立体景观设计，墙面和屋顶都要进行绿化，从设计之初的规划阶段就要纳入考虑，以便最终形成一个和谐的整体。这其中不仅涉及植被设计，还有灌溉系统和后期维护的问题，确保景观的可持续性。在现代世界中，我们会看到越来越多的绿墙和屋顶花园。立体景观设计已经从最初的装饰性绿墙发展为建筑和整体规划的一部分。设计时也要从整体上做全局考虑。

《立体景观》一书从立体景观设计的两大元素着手——绿墙与屋顶花园。除了展示世界各地最新的景观设计作品之外，本书还涉及立体绿化的技术方面，包括植物选择、灌溉系统和植被维护等。景观的设计和维护是确保建筑的美观与功能的完善的关键。

CONTENTS

Bella Sky Hotel

贝拉天空酒店

Location:
Copenhagen, Denmark
Architect:
Deichmann Planter ApS.
Photographer:
Søren Johansen & Adam Mørk
Area:
200 sqm

项目地点：
丹麦，哥本哈根
建筑设计：
丹麦戴希曼景观事务所
摄影师：
索伦・约翰森、亚当・默克
面积：
200平方米

Project description:

The opening of Scandinavia's largest hotel symbolises not only the identity of the modern Ørestad area of Copenhagen, but also marks the Capital's increasing importance on the international convention and congress scene. Visually the hotel will be characterised by the two leaning towers, creating a spectacular and unspoiled view over the green meadows, the sea and the Copenhagen skyline. The sculptural profile and visibility of the hotel from the various corners of Copenhagen make it an architectural landmark which will draw many visitors to the Ørestad neighbourhood.

The unique design creates diversity at almost every floor level and thereby challenges the scopes for conventional construction. The complexity of Bella Sky reflects not only the great engineering and constructional achievements, but also the architectural ambition to create a unique and personal hotel experience. Top twist of one of the towers decreases problems with turbulence in the flat, windy landscape, and a ground floor twist of the other makes room for a clearly indicated entrance. The hotel lobby is merged into the existing entrance lobby of the Bella Centre, making the hotel a true integrated part of the complex.

The green wall at Bella Sky consists of 4,250 plants. The plant walls are sprinkled automatically by an irrigation system, which gives the wall the correct amount of water and fertiliser. The green wall is part of Bella Centre's and Bella Sky Comwell's CSR policy to make the indoor climate more clean and improve the well-being of guests and staff.

The 196-square-metre wall is divided between two floors and fitted on concrete walls by five radius. The green wall consists of the following plants: Agleonema maria, Asplenium antigum, Chlorophytum green, Davalia, Platycerium bifurcatum, Scindapsus pictus, and Spathiphyllum. The four-star Bella Hotel will accommodate all modern conveniences in 814 rooms, 32 conference rooms, 3 restaurants, a sky bar, a wellness centre, etc.

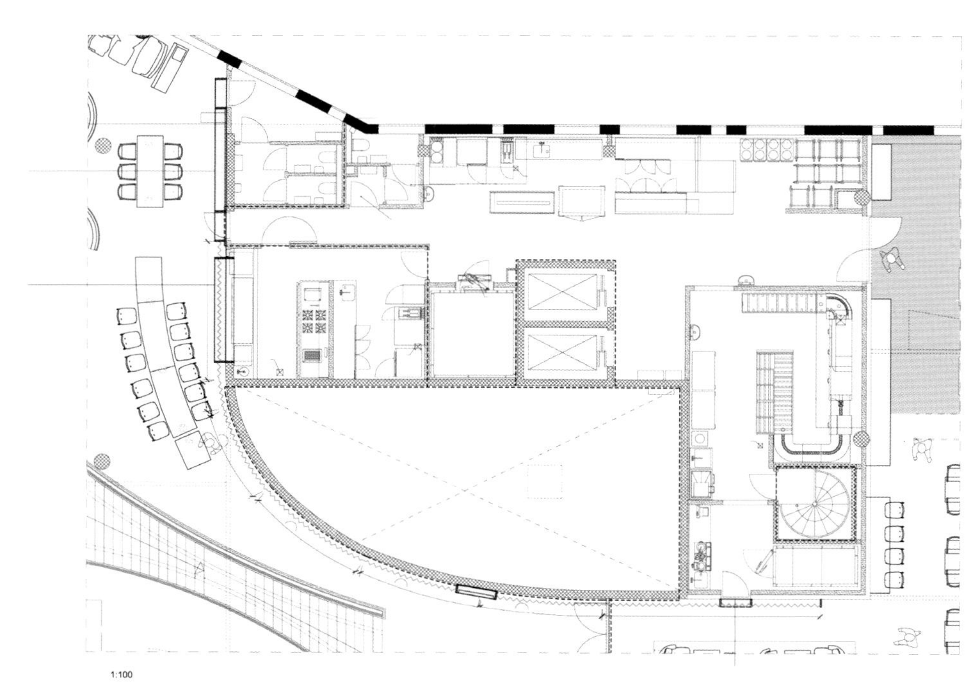

PLAN 平面图

贝拉天空酒店是斯堪的纳维亚半岛最大的酒店。这家酒店不仅代表了哥本哈根欧瑞斯塔区的现代形象，而且象征了丹麦首都在国际舞台上日益重要的地位（许多国际会议在这里举行）。这家酒店以两栋斜塔建筑著称，从这里可以眺望外面绿油油的草坪、一望无际的大海和天空，风景十分迷人。酒店的建筑造型非常别致，是哥本哈根的地标，将更多游客吸引到欧瑞斯塔区。

独特的建筑设计让几乎每个楼层都各不相同，也因此给施工带来不小的挑战。贝拉天空酒店的建筑复杂性不仅反映了建筑工程与施工上的成就，而且也体现了建筑师希望住客享受独特的、人性化的酒店体验的良苦用心。其中一栋斜塔顶部齐平，方便进行屋顶绿化；另一栋斜塔的底层打造了十分醒目的入口空间。酒店大堂与贝拉中心原来的入口大厅融为一体，让两座建筑真正合而为一。

贝拉天空酒店的绿墙采用了4250株植物。植物采用先进的自动喷洒灌溉系统，能够精确控制水量和肥料。绿墙的设计反映了贝拉中心和贝拉天空酒店的企业社会责任，目的是打造更加健康的室内气候，有益于酒店客人和工作人员的身体健康。

绿墙面积为196平方米，覆盖两个楼层，里面是混凝土墙面。主要植物有：万年青、铁角蕨、吊兰、狼尾蕨、二歧鹿角蕨、星点藤、白鹤芋等。贝拉天空酒店是一家四星级酒店，有814间客房、32间会议室和3个餐厅，此外还有空中酒吧和健身中心等，配备了现代化的配套设施。

NOTES

1. All dimensions in millimetre
2. Wood
   W5: Oak Solid
   W6: Oak veneer, to match the solid oak in colour, finish and grain direction
3. Materials
   05: Brushed Stainless Steel

备注：

1. 所有单位为毫米
2. 木材
   W5：实心橡木
   W6：橡木胶合板（颜色、表面材料和纹理都与实心橡木相匹配）
3. 材料
   05：拉丝不锈钢

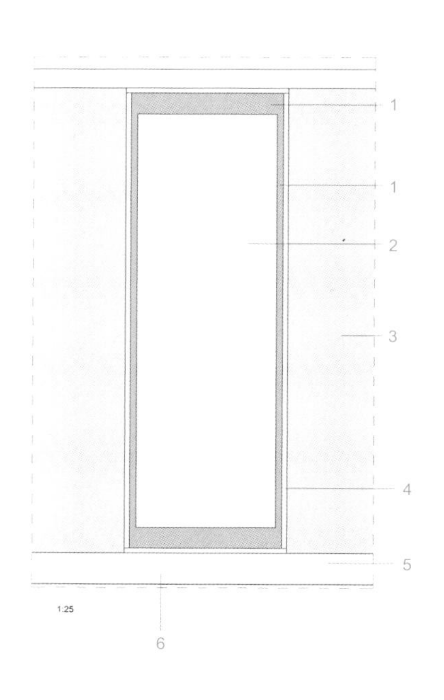

PLAN

1. If TV screen needs air circulation this surface to be a perforated metal sheet lacquered in colour to be matched with TV frame
2. TV Screen
3. Plant wall
4. W5 or W6
5. 05
6. General height of skirting board in the hotel: 100mm as per 3XN drawings

平面图

1. 如果电视屏幕需要散热，这个表面可以安装穿孔金属板（颜色与电视屏幕的边缘相匹配）
2. 电视屏幕
3. 绿墙
4. W5或W6
5. 05
6. 酒店内的标准踢脚板高度：100毫米

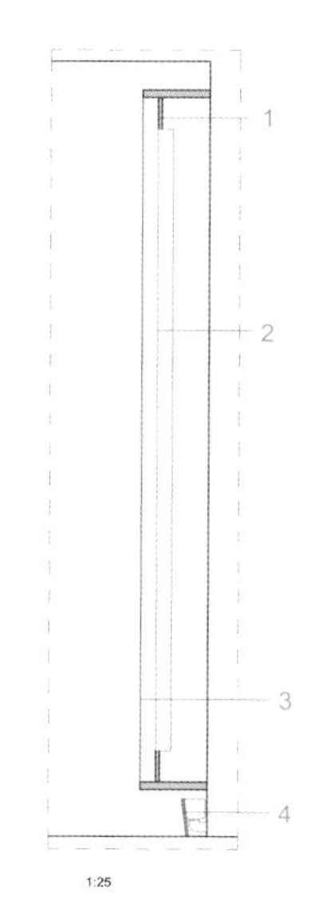

SECTION

1. If TV screen needs air circulation this surface to be a perforated metal sheet lacquered in colour to be matched with TV frame
2. TV Screen
3. W5 or W6
4. 05

剖面图

1. 如果电视屏幕需要散热，这个表面可以安装穿孔金属板（颜色与电视屏幕的边缘相匹配）
2. 电视屏幕
3. W5或W6
4. 05

PLAN
1. Plant wall
2. W5 or W6
3. 05

平面图
1. 绿墙
2. W5或W6
3. 05

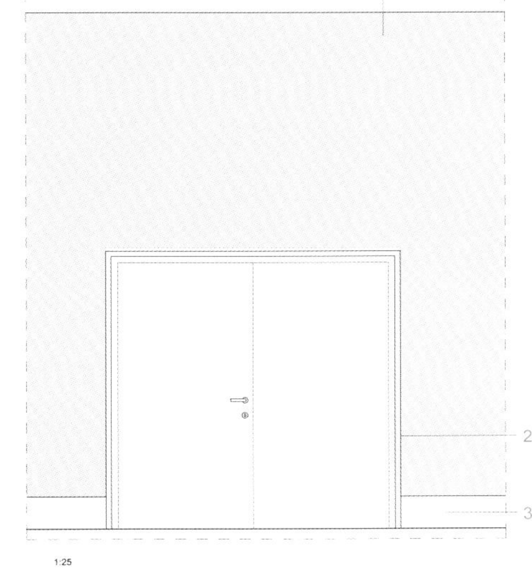

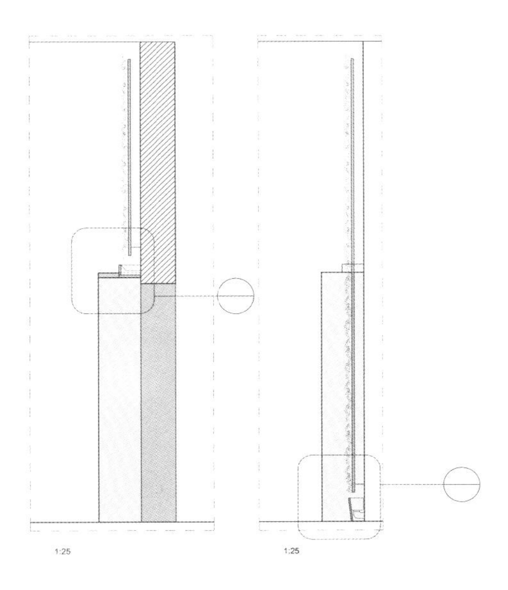

SECTION 剖面图

DETAIL 01
1. Drainage of plant wall
2. Frame around door

细部详图-01
1. 绿墙排水系统
2. 门四周的框架结构

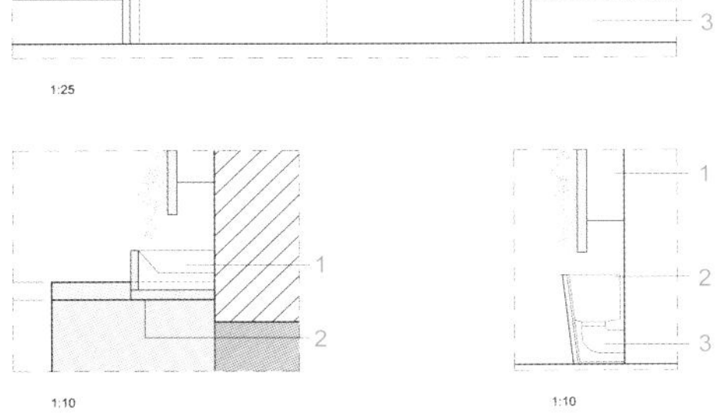

DETAIL 02
1. Distance from wall
2. Drainage of plant wall
3. General height of skirting board in the hotel: 100mm as per 3XN drawings

细部详图-02
1. 到墙壁的距离
2. 绿墙排水系统
3. 酒店内的标准踢脚板高度：100毫米

Britomart East Green Wall

布里托马特绿墙

Completion date:
2011
Location:
Auckland, New Zealand
Landscape architect:
Natural Habitats
Photographer:
Natural Habitats
Area:
120 sqm

竣工时间：
2011年
项目地点：
新西兰，奥克兰
景观设计：
"自然栖息地"景观事务所
摄影师：
"自然栖息地"景观事务所
面积：
120平方米

Project Description:

- Largest green wall in New Zealand
- Made from 60 custom-made panels

These eco-artworks are the focal point of the atrium on Takutai in the Britomart precinct. The living walls line the east and west end walls and integrate seamlessly into the building fabric. The introduction of these vertical gardens has improved air quality, acoustics and thermal performance. They provide a place for natural ecology in the commercial environment and give office occupants a sense of connection with the natural environment.

The green walls feature a custom-designed planting palette that has a combination of native and exotic species chosen for their low light and maintenance requirements. A collection of flowering plants provide seasonal variety, and have been specially chosen to ensure falling drifts of petals don't stain the steel grey tiles below. The overall composition was influenced by the shadows that fall on the wall during the day, with repetition of planting patterns loosely referencing those found in traditional Maori carvings.

As the walls are only 120mm deep, containing an inert medium as opposed to soil, they can be easily fixed into the atrium's existing stud pattern. The walls' 60 custom-made panels had to be carefully installed by Natural Habitats using a building maintenance unit and abseiling equipment. They are an ideal lightweight solution to the question of how to differentiate the complex's interior aesthetics.

"The green wall has been the single most outstanding feature of our new work environment. It causes people to stop and take time out to look at the detail of it, which is extremely important in an otherwise rushed corporate environment." —OFFICE WORKER

EXIT
SOUTHERN CROSS HEALTH SOCIETY RECEPTION →
WESTPAC RECEPTION ←

WESTERN ELEVATION
1. Louvre panel
2. Timber panel
3. Light level 2300
4. Light level 2200
5. Light level 2500
6. Light level 3000
7. Light level 4000

西侧立面图
1. 遮阳板
2. 木板
3. 照明强度2300
4. 照明强度2200
5. 照明强度2500
6. 照明强度3000
7. 照明强度4000

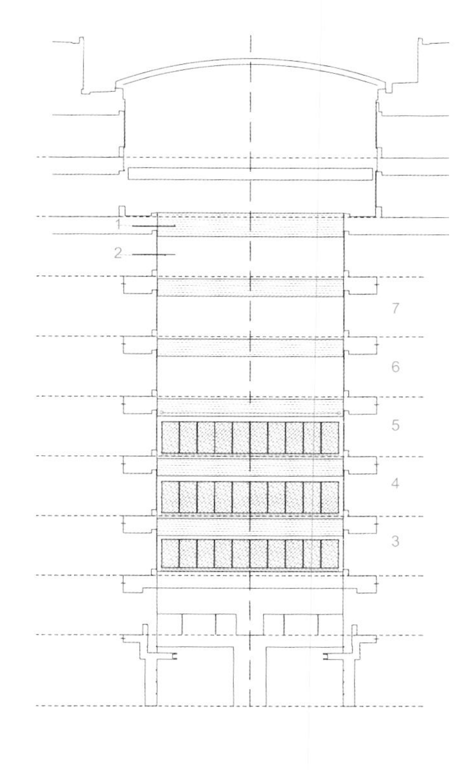

EASTERN ELEVATION
1. Louvre panel
2. Timber panel
3. Light level 4000
4. Light level 4200
5. Light level 5400
6. Light level 5700
7. Light level 8500

东侧立面图
1. 遮阳板
2. 木板
3. 照明强度4000
4. 照明强度4200
5. 照明强度5400
6. 照明强度5700
7. 照明强度8500

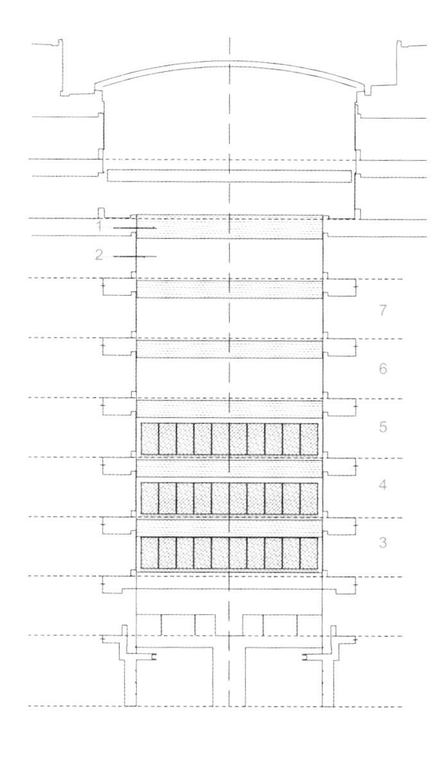

– 新西兰体量最大的绿墙
– 采用60块专用板材

PANTEL ELEVATION
1. Panel
2. Cutter fixed to bar

墙板立面图
1. 墙板
2. 与格条相连

本案中的绿墙可以说是“生态艺术品”。绿墙位于奥克兰布里托马特区一栋商业写字楼的中庭里（东侧和西侧），是整个空间的焦点。绿墙已经与建筑结构完美地融为一体。“垂直花园”的存在改善了空气质量，有助于吸音降噪和隔热保温。绿墙在商业环境中营造出自然生态系统，让在大楼中上班的员工感觉身处大自然当中一般。

绿墙上种植的植被采用本地和外来植物搭配，设计师挑选的都是无需过多光照、维护方便的植物。其中包含多种开花植物，确保绿墙根据季节的变化呈现出不一样的景象。此外，植物的选择还特别注意到不要让掉落的花瓣沾污下面的青灰色的瓷砖。绿墙最终呈现出来的整体形象取决于投射在上面的光与影，一天之中会有很大不同。植被形成的图案有点儿像毛利人的传统雕刻。

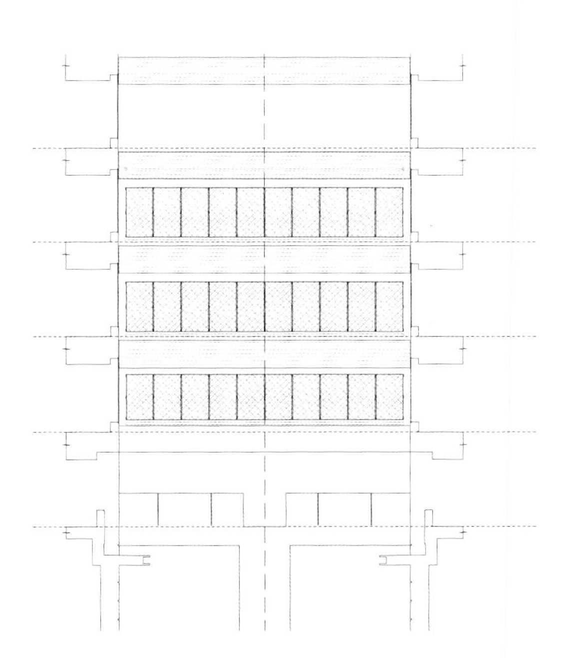

GREENWALL ELEVATION
绿墙立面图

Coffe Shop @ Athens

雅典咖啡厅

Completion date:
2013
Location:
Athens, Greece
Architect:
314 Architecture Studio
Photographer:
Panayiotis Vumbakis
Area:
300 sqm

竣工时间：
2013年
项目地点：
希腊，雅典
建筑设计：
314建筑设计工作室
摄影师：
帕纳约蒂斯・万巴基斯
面积：
300平方米

Project Description:

314 Architecture Studio designed this Coffee Shop in the city of Athens, Greece in 2013 and it offers a garden-like design that combines the modern trend with the natural beauty.

Spreading over 300 square metres, the green walls it features are placed in semi-opened spaces and thus it creates the impression of being a part of a garden. The roof has several openings which provide natural light and they can be covered with metal shades during the summer, when the sun is too powerful.

The concept was to create a chilled environment that gives the impression of being in a garden. For this reason green walls were installed in two semi-opened spaces. These spaces get sunlight from the openings in the roof and in summer months the metal shades protect from direct sunlight. The architects used modern furniture such as LC2 sofas to create a retro feeling. The materials used are mostly natural ones such as wood, cement and rusted metal with concrete for the industrial style sculpture which was created by Gianni Aspras, giving an industrial style in the coffee shop. Other artworks also adorn the coffee shop such as pictures of street art or graffiti and the furniture is modern and brings the place up to date. The building, an old industrial building, got a bioclimatic shading system with perforated metal panels.

2013年，314建筑设计工作室设计了这间雅典咖啡厅。设计师将现代的设计手法与大自然的美完美结合，打造了如花园一般的室内环境。

绿墙是这家咖啡厅内最大的特色，绿墙面积达到300多平方米。室内采用半开放式布局，使得绿墙仿佛成为花园中的一部分。天花板上有几个天窗，让自然光线照射进来。夏天阳光过于炽烈时，可以用金属遮阳板进行遮挡。

设计理念是打造凉爽的室内环境，使人感觉如同置身花园中一般。为了实现这个目标，设计师在两个半开放式空间中设计了绿墙。这两个空间的屋顶上都留有天窗，保证植物的自然采光，夏天可以用金属遮阳板保护植物不受阳光直射。设计师采用现代家具，如柯布西耶沙发（LC2），却营造出复古的感觉。使用的材料大部分是自然材料，如木材、水泥和回收利用的金属等。詹尼·阿斯普拉斯设计的工业风雕塑采用混凝土制成，赋予整间咖啡厅一种工业风格。店内还采用了其他艺术装饰品，包括挂在墙上的街头艺术绘画和涂鸦艺术。店内家具都很现代，让店内空间显得很新潮。这家咖啡厅所在的建筑是一栋古老的工业建筑，板材是穿孔金属板，本身就能营造出微型生物气候。

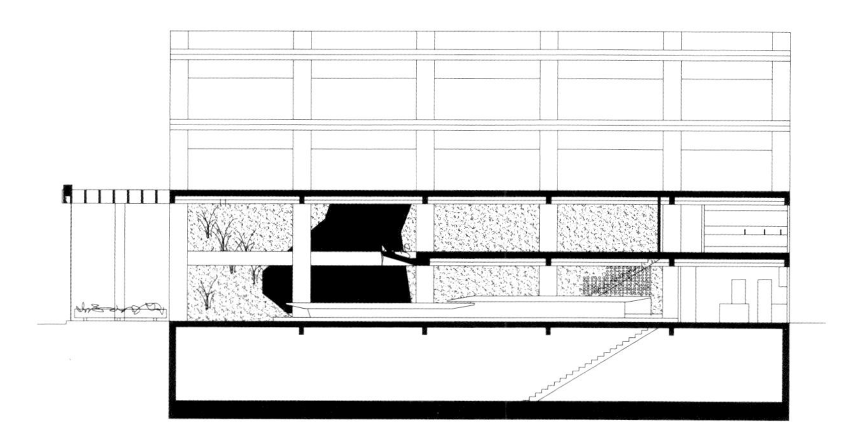

SECTION 剖面图

Freehills Offices

史密夫·菲尔律师事务所

Completion date:
2013
Location:
Sydney, Australia
Architect:
BVN
Landscape architect:
360 Degrees
Design & Construct:
Fytogreen
Photographer:
Stuart Tyler
Area:
60 sqm

竣工时间：
2013年
项目地点：
澳大利亚，悉尼
建筑设计：
BVN建筑事务所
景观设计：
"360度"景观事务所
设计&施工：
菲多格林公司
摄影师：
斯图尔特·泰勒
面积：
60平方米

Project description:

Freehills are relocating thier Sydney office to the newly constructed 161 Castlereagh Street commercial building, constructed by Grocon Developments NSW Pty, which will have a total NLA of 54,000 square metres over 37 office levels, with retail and apartment space.

The Freehills legal offices project involved five separate vertical gardens located indoors on levels 26, 30, 33, 35 and 38 of the recently constructed inner city corporate office, totalling 60 sqm.

Fytogreen worked with the Architects BVN for the initial design concept and ensuring all the host wall and services enable the gardens to survive. For the aesthetic of the garden design, Fytogreen worked with Landscape Architects 360 Degrees so the "look" matches the Ecological Sustainability. Installation was done separately for each level and each vertical garden has its own control system.

The planting plan for each level has an individual look to compliment the interior design of that level. On level 26, 30, and 38 the vertical garden acts as a backdrop to the staircase landing area. On level 33 the vertical garden is in a waiting space with TV viewing and city views. On level 35 the garden is mounted on an "S" curved host wall creating a backdrop to the reception area.

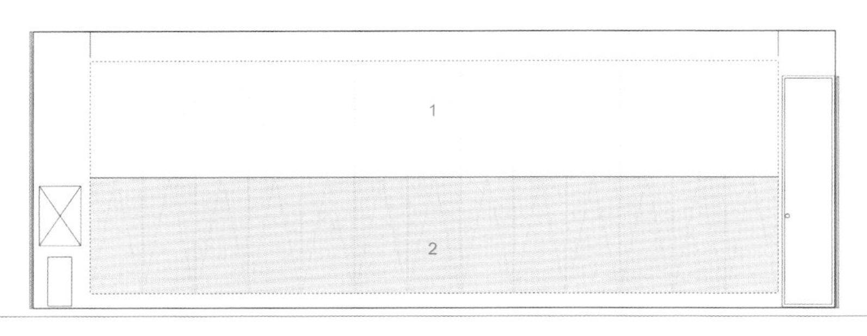

LEVEL 26 - GREEN WALL PLANTING PLAN

1. Planting Palette 1
   Height: Tall
   Form: Hanging
   Foliage: Fine/soft
   Colour: Lime/light green
   Type: Mixed species/ferns
   e.g. Davallia, Adiantum, Nephrolepis
2. Planting Palette 2
   Height: Low to medium
   Form: Upright
   Foliage: Broad leaf
   Colour: Mid to dark green
   Type: Mixed species
   e.g. Epipremnum, Peperomia, Philodendron

26层绿墙种植平面图

1. 种植法1
   高度：较高
   形态：悬垂
   叶片：细密、柔软
   颜色：石灰绿、浅绿
   类型：多品种混合（蕨类）
   品种举例：兔脚蕨、铁线蕨、肾蕨
2. 种植法2
   高度：中低
   形态：直立生长
   叶片：大叶
   颜色：中绿到深绿
   类型：多品种混合
   品种举例：黄金葛、皱叶椒草、蔓绿绒

LEVEL 26 - GREENWALL PLANTING MONTAGE
26层绿墙植被效果图

NOTE

Light, Temperature and Humidity

This indoor vertical garden design assumes a minimum of 3000 lux light exposure at every point on the host wall surface. Air temperature is to be predominantly maintained at 15 ~ 25 dCelsius. No conditioned air is to be directly vented on the plants, which would decrease humidity on the panel surface and stress foliage.

Aesthetics

The aethetis is predominantly darker green, broad-leaf, foliage to the lower half of the garden, with finer textured, lighter green, foliage to the top half. Although Philodendron scandens is arrayed within the lighter green top half. It is functionally positioned to provide foliage coverage beneath Adiantum 'Fritz Luthii' foliage, which will in time expand outwards and cascade downwards, obscuring the Philodendron. This treatment is required as self-shading from Adiantum, needs to be countered by a flat-growing, extreme low-light tolerant species, which P. scandens is.

备注

光照、温度和湿度：这个室内垂直花园的设计能够确保墙面上的每个点都有至少3000勒克斯的光照。空气温度一般控制在15～25摄氏度。植物不直接用空调吹风，那样会导致种植槽和叶片的湿度降低。

外观：绿墙的外观可以分为两个部分：下半部分是大叶植物，呈现深绿色；上半部分是小叶的密集植物，呈现浅绿色。尽管蔓绿绒种植在浅绿色的上半部分，但是栽种的位置能够确保其为铁线蕨的叶片提供绿色背景。随着时间流逝，铁线蕨会大量生长，倾泻而下，逐渐遮挡住蔓绿绒。采用这种处理是因为铁线蕨需要一种在平面上生长的、耐极弱光的植被作为底衬，而蔓绿绒正是这样的植物。

VERTICAL GARDEN LEVEL 26 PLANTING PLAN

Species Selection

A = Adiantum 'Fritz luthii'	x26
Am = Aglaonema 'Marie'	x14
C = Clivia miniata (Belgian Hybrid)	x17
D = Davalia fejiensis	x124
P = Philodendron scandens	x154
S = Spathiphyllum 'Power Petite'	x127
TOTAL	462

26层垂直花园平面图

植物品种
A = 铁线蕨（26株）
Am = 粗肋草（14株）
C = 君子兰（比利时杂交品种，17株）
D = 兔脚蕨（124株）
P = 蔓绿绒（154株）
S = 白鹤芋（127株）
共计：462株

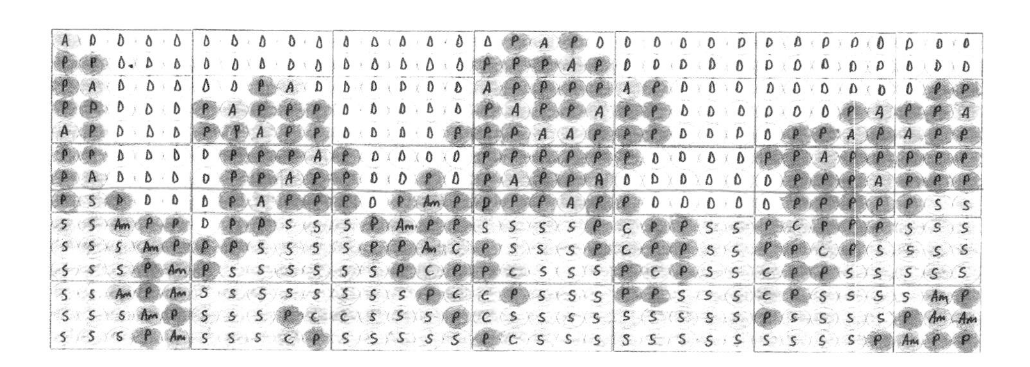

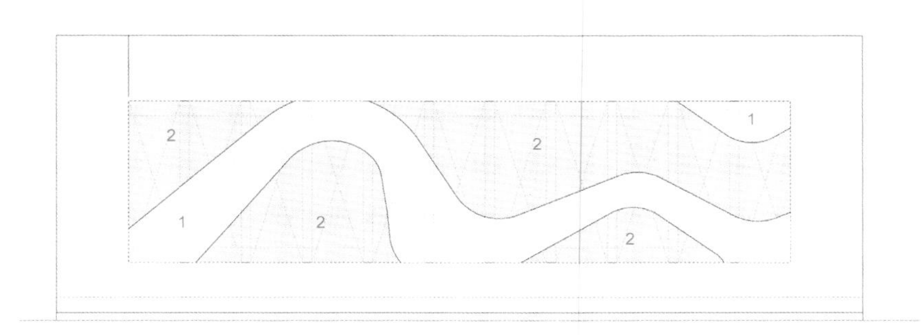

LEVEL 30 - GREEN WALL PLANTING PLAN

1. Planting Palette 3
   Height: Tall
   Form: Flowingv
   Foliage: Fine/soft
   Colour: Lime/light green
   Type: Mixed species
   e.g. Chlorophytum, Davallia, Nephrolepis
2. Planting Palette 4
   Height: Medium consistent
   Form: Upright shrub
   Foliage: Broad leaf
   Colour: Mid to dark green
   Type: Mixed species
   e.g. Epipremnum, Peperomia, Philodendron

30层绿墙种植平面图

1. 种植法3
   高度：较高
   形态：流线型
   叶片：细密、柔软
   颜色：石灰绿、浅绿
   类型：多品种混合
   品种举例：吊兰、兔脚蕨、肾蕨
2. 种植法4
   高度：始终保持中等高度
   形态：直立生长的灌木
   叶片：大叶
   颜色：中绿到深绿
   类型：多品种混合
   品种举例：黄金葛、皱叶椒草、蔓绿绒

LEVEL 30 - GREEN WALL PLANTING MONTAGE
30层绿墙植被效果图

NOTE

Light, Temperature and Humidity refer to level 26 on page26

Aesthetics: Species compatibilities have been acutely assessed, rendering the above design, based on the aesthetic concept provided. Epipremnum has been removed, due to currently supply only offering variously variegated cultivars. The use of variegation would detract from the provided concept aesthetic, and therefore Spathiphyllum and Philodendron 'Super Atom' have been substituted in as alternative dark green foliaged, compatible species. Chlorohpytum has been removed from within the fern swathe, due to its growth form increasing shading, thereby rendering it uncompatible.

备注

光照、温度和湿度见26页26层垂直花园平面图。

外观：设计师充分考虑了植物品种相容性，在此基础上营造出如26层垂直花园一样的外观。原来的黄金葛没再使用，避免色彩过于斑驳。斑驳的色彩会将绿墙上下两半的外观结构打乱，所以设计师用白鹤芋和蔓绿绒取而代之，因为这两种植物叶片呈深绿色，与绿墙整体外观相容。蕨类种植区里原有的吊兰也没再用，因为过度生长的吊兰会影响其他植物的光照，让整体绿墙显得不相容。

VERTICAL GARDEN LEVEL 30 PLANTING PLAN

Species Selection

D = Davalia fejiensis	x124
N = Nephrolepis cordifolia 'Duffy'	x9
P = Philodendron scandens	x45
Pe = Peperomia caperata 'Emerald Ripple'	x54
S = Spathiphyllum 'Power Petite'	x53
TOTAL	264

30层垂直花园平面图

植物品种
D = 兔脚蕨（124株）
N = 肾蕨（9株）
P =蔓绿绒（45株）
Pe = 皱叶椒草（33株）
S = 白鹤芋（53株）
共计：264株

S	S	S	P	P	D	D	D	D	D	D	P	S	S	Pe	Pe	Pe	Pe	Pe	Pe	S	P	P	P	P	S	S	S	D	D	D	D	D
S	S	P	P	D	D	D	D	D	D	D	D	P	S	Pe	Pe	Pe	Pe	Pe	S	S	P	P	P	P	P	S	S	S	D	D	D	D
S	P	P	D	D	D	D	D	D	D	D	D	D	S	Pe	Pe	Pe	S	S	S	P	D	D	D	D	P	P	S	S	S	D	D	S
P	P	D	D	D	D	Pe	Pe	Pe	Pe	D	D	D	P	S	S	S	S	P	P	D	D	D	D	D	D	P	P	P	S	S	S	P
P	D	D	D	D	D	Pe	Pe	Pe	S	D	D	D	P	P	P	P	P	D	D	D	D	D	D	D	D	D	P	P	P	P	P	P
N	D	D	D	D	Pe	Pe	Pe	S	S	S	D	D	D	D	D	D	D	D	D	D	S	Pe	Pe	D	D	D	D	D	D	D	D	D
N	D	D	D	D	Pe	S	S	P	S	S	D	D	N	D	D	D	D	D	D	S	S	Pe	Pe	Pe	D	D	D	D	D	N	D	D
N	N	D	D	S	S	S	P	P	P	S	S	D	N	N	D	D	D	D	S	S	S	S	Pe	Pe	Pe	D	D	D	D	N	D	D

LEVEL 33 - GREENWALL PLANTING MONTAGE
33层绿墙植被效果图

LEVEL 33 - GREENWALL PLANTING PLAN
1. Planting Palette 5
   Height: Mixed
   Form: Mixed
   Foliage: Mixed- fine, broad, strappy
   Colour: Predominantly lime green/light green
   Type: Mixed 'Rainforest' species
   e.g Selaginella, Davallia, Adiantum, Huperzia, Nephrolepis, Asplenium

33层绿墙种植平面图
1. 种植法5
   高度：混合型
   形态：混合型
   叶片：混合型（细密、大叶、长叶）
   颜色：以石灰绿和浅绿为主
   类型：多“雨林”品种混合
   品种举例：卷柏、兔脚蕨、石杉、肾蕨、铁角蕨

澳大利亚知名的史密夫・菲尔律师事务所将其悉尼办公地点搬到繁华的卡苏里大街161号新建的一栋商业大厦中。这栋大楼由新南威尔士的澳洲顶级承建商格罗康公司建造，净租赁面积54,000平方米，除了37层写字间之外，还有公寓和零售空间。

史密夫・菲尔律师事务所的室内景观设计包括5面绿墙，或者说5个“垂直花园”，分别位于26、30、33、35、38这几个楼层上，绿化面积共计60平方米。

菲多格林公司在设计之初就与BVN建筑事务所紧密合作，从最初的设计理念开始，确保绿墙设备能够满足植物的生长条件。景观设计由“360度”景观事务所负责，菲多格林公司也与之开展合作，让绿墙最终呈现出“生态园”一般的繁茂景象。各个楼层的绿墙分别施工，每个“垂直花园”有自己的控制系统。

每个楼层上的绿墙，植被设计各不相同，都与相应楼层的室内设计相一致。26、30和38这几个楼层，绿墙作为楼梯的背景出现。33层的“垂直花园”是等候区，配有电视机，从窗口能眺望悉尼风景。35层的绿墙呈现出“S”形的曲线，是接待区的背景墙。

VERTICAL GARDEN LEVEL 33 PLANTING PLAN

Species Selection

A = Adiantum	x5
Ch = Chamaedorea elegans	x13
C = Chlorophytum comosum 'green'	x5
D = Davalia fejiensis	x124
E = Epipremnum 'Green & Gold'	x35
Eg = Epipremnum 'Goldilocks'	x11
Nd = Nephrolepis cordifolia 'Duffy'	x7
Ne = Nematanthus glabra	x8
Ng = Neomarica gracilis	x8
Pe = Peperomia caperata 'Emerald Ripple'	x48
S = Spathiphyllum 'Power Petite'	x54
Sp = Syngonium 'Pixie'	x4
Ss = Solerolia soleirolii	x16
TOTAL	284

33层垂直花园平面图

植物品种
A = 铁线蕨（5株）
Ch = 袖珍椰子（13株）
C = 吊兰（5株）
D = 兔脚蕨（124株）
E = "绿金"黄金葛（35株）
Eg = "金凤"黄金葛（11株）
Nd = 肾蕨（7株）
Ne = 袋鼠花（8株）
Ng = 巴西鸢尾（8株）
Pe = 皱叶椒草（48株）
S = 白鹤芋（54株）
Sp = 合果芋（4株）
Ss = 绿珠草（16株）
共计：284株

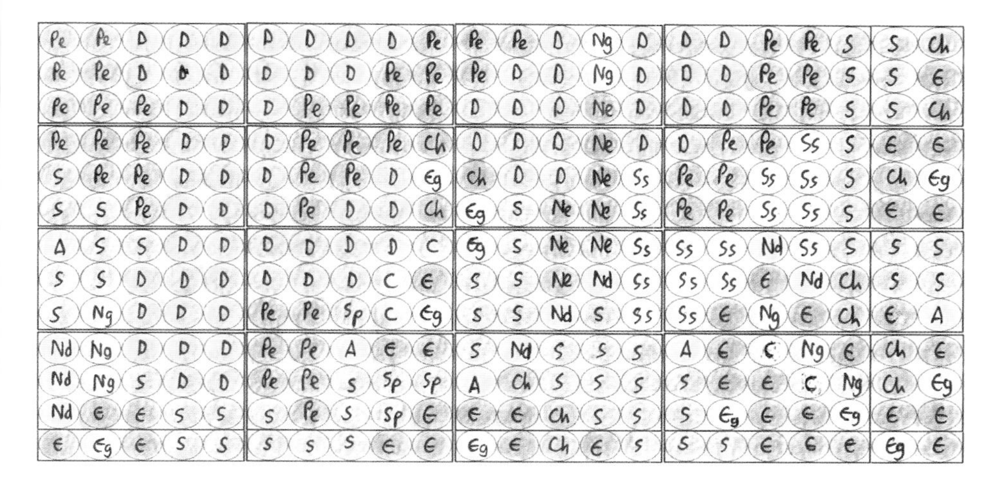

NOTE

Light, Temperature and Humidity refer to level 26 on page26

Aesthetics

The above colours do not represent foliage colour, rather species groupings per type. Species compatibilities have been acutely assessed, rendering the above design, based on the aesthetic concept provided.

Lighter green species have been used, where possible, pending availability and suitability. Where only darker green foliage species have been available, these have been inserted in smaller groupings. The darker green foliaged species are minimal, yet are required, to either control the growth of more aggressive species, separate species and colonise low light areas.

备注

光照、温度和湿度见26页26层垂直花园平面图。

外观：

以上颜色不代表植物叶片的颜色，而是多个品种杂糅的颜色。设计师充分考虑了植物品种相容性，在此基础上营造出如26层垂直花园一样的外观。

在兼顾可行性和适宜性的基础上，设计师尽可能采用浅绿色的植物。如果只有深绿色叶片的植物可用，则将其栽种成较小的簇。深绿色叶片的植物用的很少，但也需要，作用有三：控制某些侵略性植物的过度生长；将不同品种进行区分；栽种于弱光区。

SPECIESILLUSTRATIONS 植物样图

Adiantum 铁线蕨	Chamaedorea 袖珍椰子	Chlorophytum 吊兰	Davalia 兔脚蕨	Epi 'Green & Gold' "绿金"黄金葛
Epi 'Goldilocks' "金凤"黄金葛	Nephro 'Duffy 肾蕨	Nematanthus 袋鼠花	Neomarica 巴西鸢尾	Peperomia 皱叶椒草
Spathiphyllum 白鹤芋	Syngonium 合果芋	Soleirolia 绿珠草		

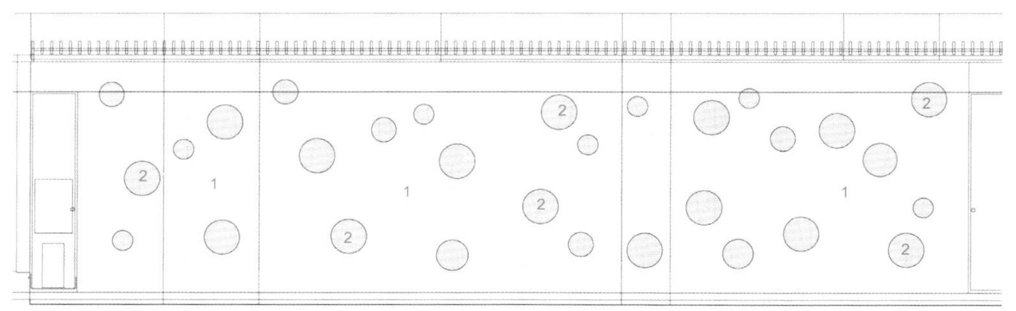

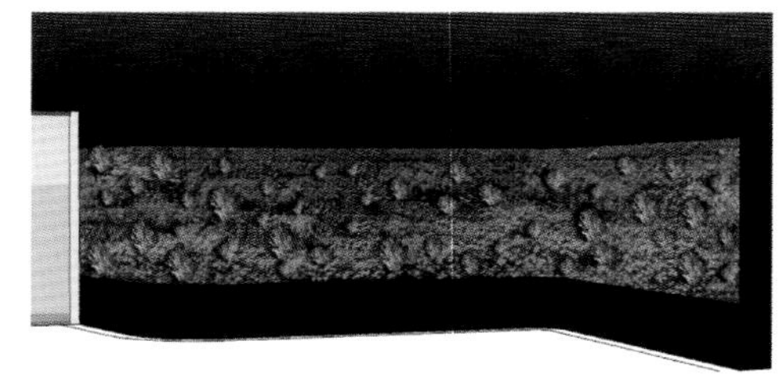

LEVEL 35 - GREEN WALL PLANTING MONTAGE
35层绿墙植被效果图

LEVEL 35 - GREEN WALL PLANTING PLAN

1. Planting Palette 6
   Height: Low
   Form: Mixed
   Foliage: Small
   Colour: Varies- predominantly mid green
   Type: Mixed species
   e.g. Peperomia, Selaginella, Epipremnum, Philodendron
2. Planting Palette 7
   Height: Tall
   Form: Clumping/Hanging
   Foliage: Strappy/fine
   Colour: Light green
   Type: Mixed species
   e.g. Chlorophytum, Clivia, Huperzia

35层绿墙种植平面图

1. 种植法6
   高度：低矮
   形态：混合型
   叶片：小叶
   颜色：多变（中绿为主）
   类型：多品种混合
   品种举例：皱叶椒草、卷柏、黄金葛、蔓绿绒
2. 种植法7
   高度：较高
   形态：聚丛、悬垂
   叶片：长叶、细密
   颜色：浅绿
   类型：多品种混合
   品种举例：吊兰、君子兰、石杉

VERTICAL GARDEN LEVEL 35 PLANTING PLAN

Species Selection

C = Chlorophytum comosum 'green'	x14
Cl = Clivia miniata	x50
E = Epipremnum 'Green & Gold'	x187
P = Philodendron scandens	x199
TOTAL	450

35层垂直花园平面图

植物品种

C = 吊兰（14株）

Cl = 君子兰（50株）

E = "绿金"黄金葛（187株）

P = 蔓绿绒（199株）

共计：450株

NOTE

Light, Temperature and Humidity refer to level 26 on page 26

备注

光照、温度和湿度见26页26层垂直花园平面图

SPECIES ILLUSTRATIONS 植物样图

Clivia miniata 君子兰 | Chlorophytum comosum 吊兰 | Epipremnum 'Green & Gold' "绿金"黄金葛 | Philodendron scandens 蔓绿绒

Level 38 38层

Hotel Novotel at Auckland Airport

奥克兰机场诺富特酒店绿墙

Completion date:
2011
Location:
Auckland, New Zealand
Landscape architect:
Natural Habitats
Photographer:
Natural Habitats
Area:
60 sqm

竣工时间：
2011年
项目地点：
新西兰，奥克兰
景观设计：
"自然栖息地"景观事务所
摄影师：
"自然栖息地"景观事务所
面积：
60平方米

Project description:

The new Hotel Novotel at Auckland Airport was designed to combine New Zealand's distinctive cultural heritage with modern architecture. The green wall, designed, built and installed by Natural Habitats, was intended to complement the hotel's distinctive New Zealand theme and act as a focus for the bar area.

The swathe of green that envelops the bar from floor to ceiling will improve the bar's indoor air quality by removing air pollutants and raising humidity levels. The 60-square-metre vertical vegetation provides both visual interest and environmental benefits, creating a more comfortable environment to work and relax in.

In keeping with the hotel's New Zealand theme, the green wall features an array of native New Zealand flora, including cascading ferns and pan-pacific creepers. The wall also has special significance to Tainui, who were a partner in the hotel's development. Nestled within the wall is Winika cunninghamii – an epiphytic orchid. This indigenous orchid grows on native Podocarpus totara trees, one of which was used to make the hull of a sacred waka (great ocean-going canoe) belonging to Tainui people. They named the canoe "te winika" after the orchid. The Novotel Hotel marks the vicinity of the landing place of the legendary Tainui Waka, in which Polynesians migrated to New Zealand approximately 800 years ago.

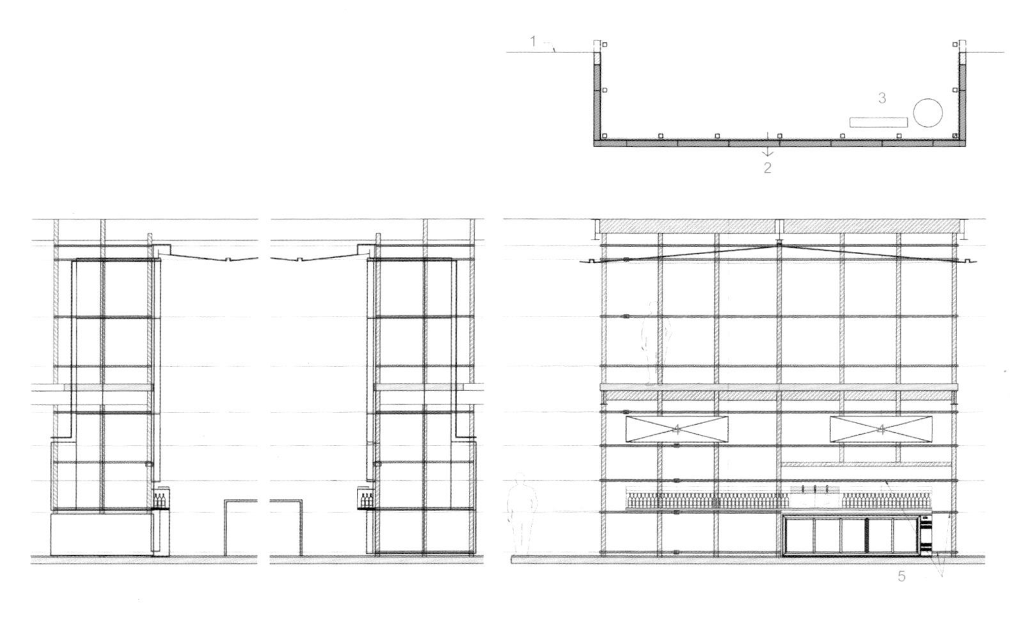

ELEVATION

1. Mezanine floor edge
2. Feed from plant room to outlets
3. Green wall plant equipment approximate location
4. Aircon
5. Note: rail support

立面图

1. 跃层楼面边缘
2. 灌溉水从种植槽的排水口流出
3. 绿墙种植设备的大概位置
4. 空调
5. 注意：扶手支撑结构

ELEVATION

1. Note: Unistrut rail extended beyond panel, TC
2. Bench to be installed post green wall installation finish
3. Note: Bench to extend to rear front wall. Bench ht reduced
4. Gutter
5. Green wall gutter below finished floor level

立面图

1.注意：扶手（Unistrut元件）要伸出支架板外
2.长凳安装在绿墙背后
3.注意：长凳一直延伸到前墙后部；长椅高度降低
4.排水槽
5.嵌入地面的绿墙排水槽

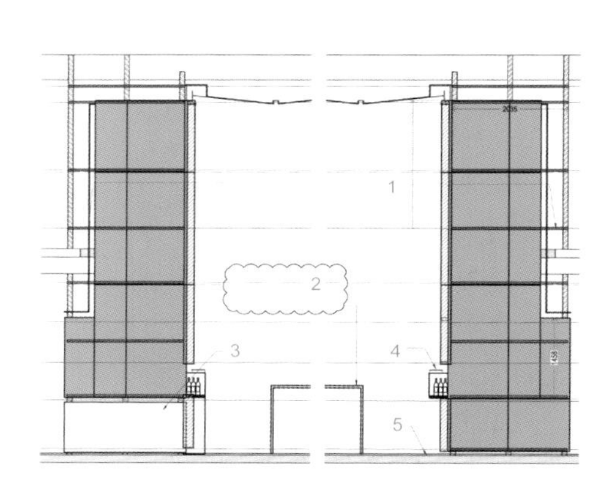

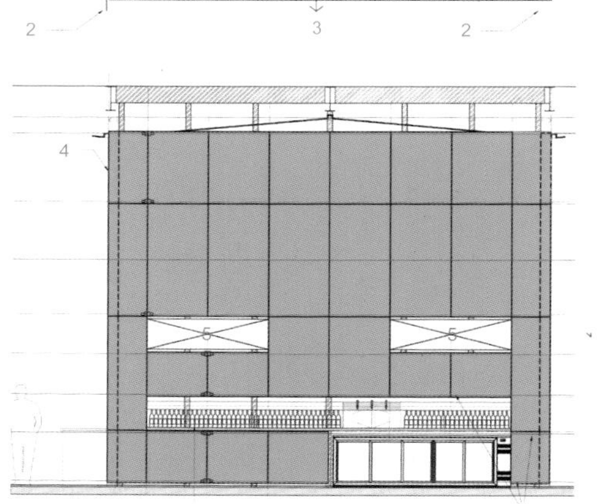

ELEVATION

1. Mezanine floor edge
2. Dimension from front of unistrut rail to grid line “6”
3. Feed from plant room to outlets
4. Green wall
5. Aircon

立面图

1. 跃层楼面边缘
2. 尺寸：从扶手（Unistrut元件）前面到坐标线“6”
3. 灌溉水从种植槽的排水口流出
4. 绿墙
5. 空调

奥克兰机场新建了一家诺富特酒店，其设计旨在将新西兰独特的文化遗产与现代建筑相结合。绿墙的设计、建造和安装全部由新西兰“自然栖息地”景观事务所操刀，目的是突出独特的新西兰特色，同时为酒吧区打造视觉焦点。

绿墙从地面一直延伸到天花板，将酒吧笼罩在绿色的氛围中。植物能够吸收空气污染物，增加空气湿度，进而改善酒吧的室内空气质量。这面绿墙面积为60平方米，既为改善室内气候带来益处，同时又美化了室内环境。不论是工作还是休闲，人们都能享受更加健康、美观的环境。

为了与这家酒店的新西兰特色更加贴合，设计师为绿墙选择了一系列新西兰本地植物，包括大面积的蕨类和太平洋地区藤蔓植物。绿墙对于泰努伊人（新西兰的一个部落）来说具有特别的意义——这家酒店的合伙人当中就有泰努伊人。绿墙中的石斛是一种附生兰花。这种石斛是本地特有的物种，长在罗汉松上，而罗汉松是泰努伊人制作“独木圣舟”的木材之一，他们用这种兰花的名字来命名这种独木舟。通过这一设计，诺富特酒店纪念了泰努伊人登陆的地方——大约800年前，波利尼西亚人就是从这里迁徙到新西兰。

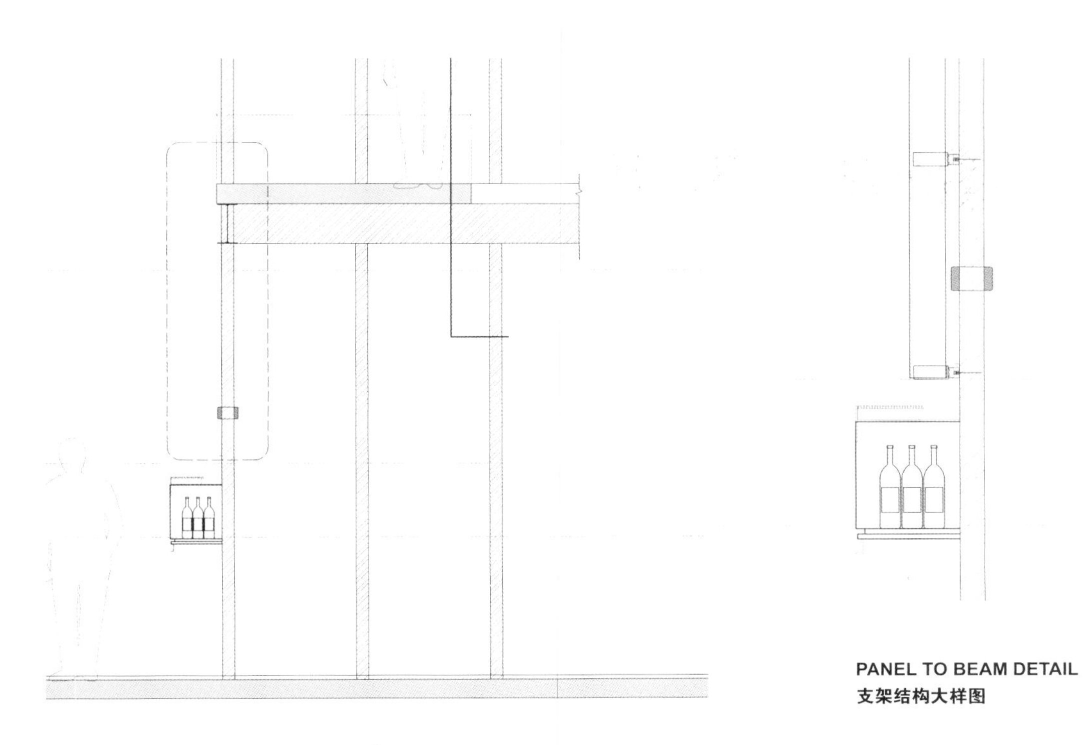

PANEL TO BEAM DETAIL
支架结构大样图

STEEL SUPPORT AS BUILT ELEVATION
支撑钢结构立面图

KAA Restaurant

KAA餐厅

Location:
São Paulo, Brazil
Architect:
Studio Arthur Casas
Photographer:
Leonardo Finotti
Area:
700 sqm

项目地点：
巴西，圣保罗
建筑设计：
亚瑟・卡萨斯设计工作室
摄影师：
莱昂纳多・费诺蒂
面积：
700平方米

Project description:

In one of the busiest streets of São Paulo the white façade of KAA restaurant hides a small oasis within the city. The plot is deep and shallow, generating a simple layout with lounge and bar at the front, tables and kitchen at the back of the restaurant. The mezzanine is occupied by toilets and a dining space.

A large pool welcomes visitors, accentuating the serenity of the ambient. The green wall with over 7,000 tropical plants dominates the side of the restaurant. A retractable translucent roof allows natural lighting. On the left side the wooden ceiling under the mezzanine contrasts with the double height in front of the green wall.

Warm and discrete materials reinforce the intimate atmosphere by using wood, ceramics and neutral colours. Most of the furniture was designed by the Studio, such as the reinterpretation of the famous Barcelona chair.

The Brazilian spirit is present by the ludic contrast between contemporary materials and local elements such as straw and indigenous artifacts. The rationality of the project is clearly represented by the rhythm in the distribution of the space, following the structure and taking advantage of the singular proportions of the building.

KAA restaurant was awarded with the Best Restaurant Design of the Year 2010 by Wallpaper Magazine.

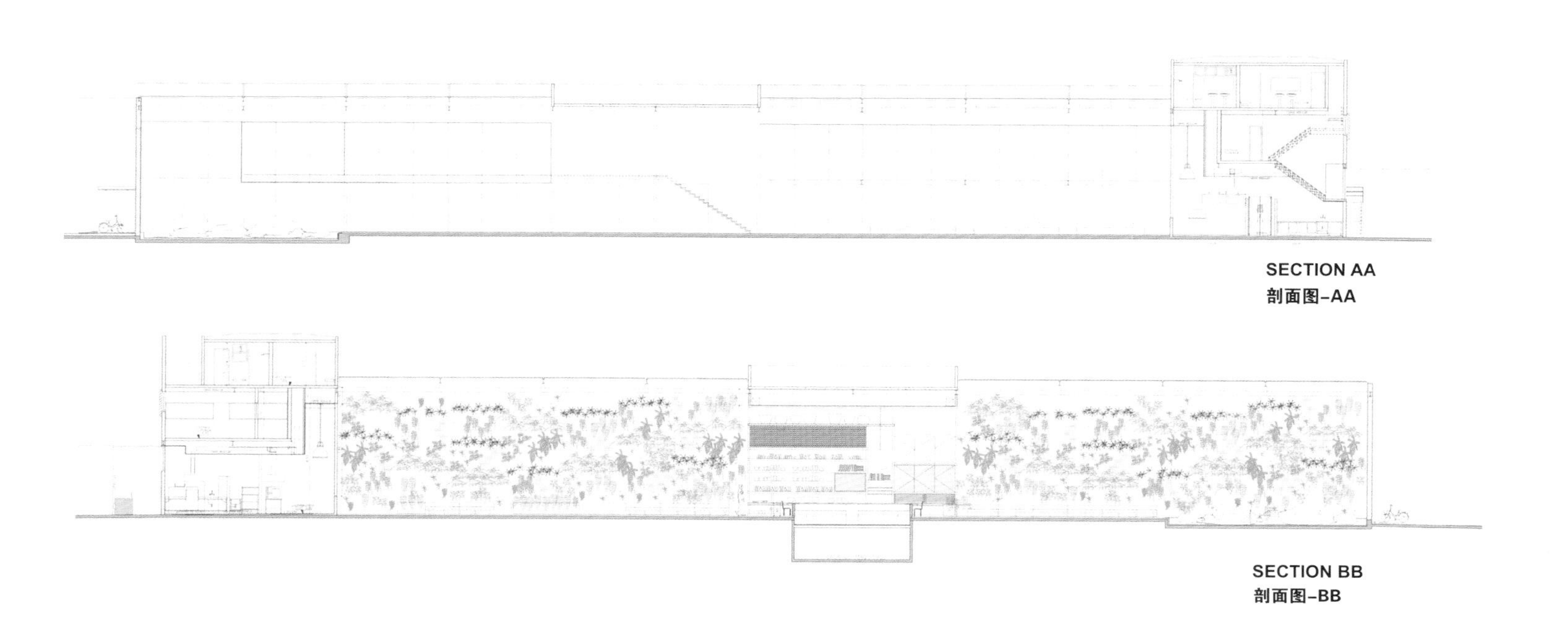

SECTION AA
剖面图–AA

SECTION BB
剖面图–BB

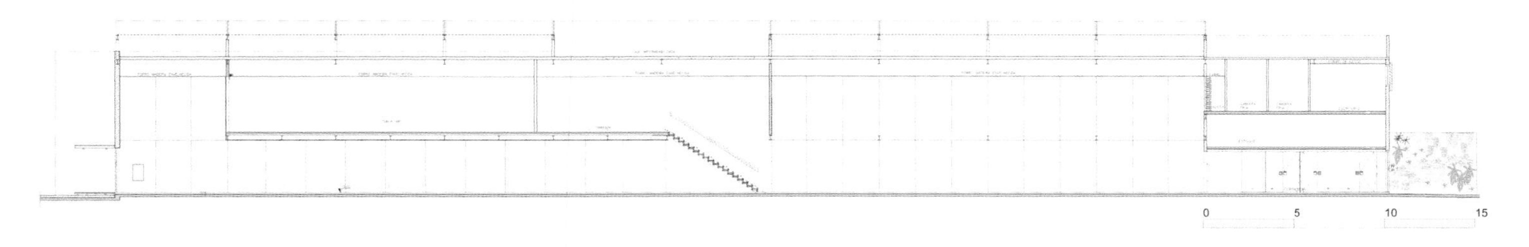

SECTION CC
剖面图-CC

KAA餐厅位于巴西圣保罗最繁华的街道上，在白色的外立面之下，隐藏着一小片绿洲。餐厅所在的地块十分狭小，所以空间布局也很简单：餐厅前面是休闲区和酒吧，后面是坐席和厨房。跃层上是卫生间和就餐区。

餐厅入口处有个巨大的水池，奠定了室内静谧的环境基调。一侧墙面进行了绿化，墙上种植着7000多种热带植物，成为整个空间的视觉焦点。半透明的屋顶可以伸缩折叠，保证了室内的自然采光。左侧，跃层下方的木质天花板与绿墙前方两层通高的开敞空间形成鲜明对照。

设计师通过使用木材、陶瓷等温暖的材料，以及中性的色彩，突出了室内温馨宜人的氛围。大部分家具都是亚瑟·卡萨斯设计工作室自己设计的，比如椅子，就是仿造著名的巴塞罗那椅。

KAA餐厅通过现代材料与代表当地传统文化的元素（如稻草和当地手工艺品）的对比使用，体现出浓郁的巴西风情。餐厅的空间布局非常合理，展现出设计师理性的一面。设计师利用建筑既定的结构和比例，营造出空间的韵律。

KAA餐厅获得了《墙纸》杂志评选的“2010年度最佳餐厅设计奖”。

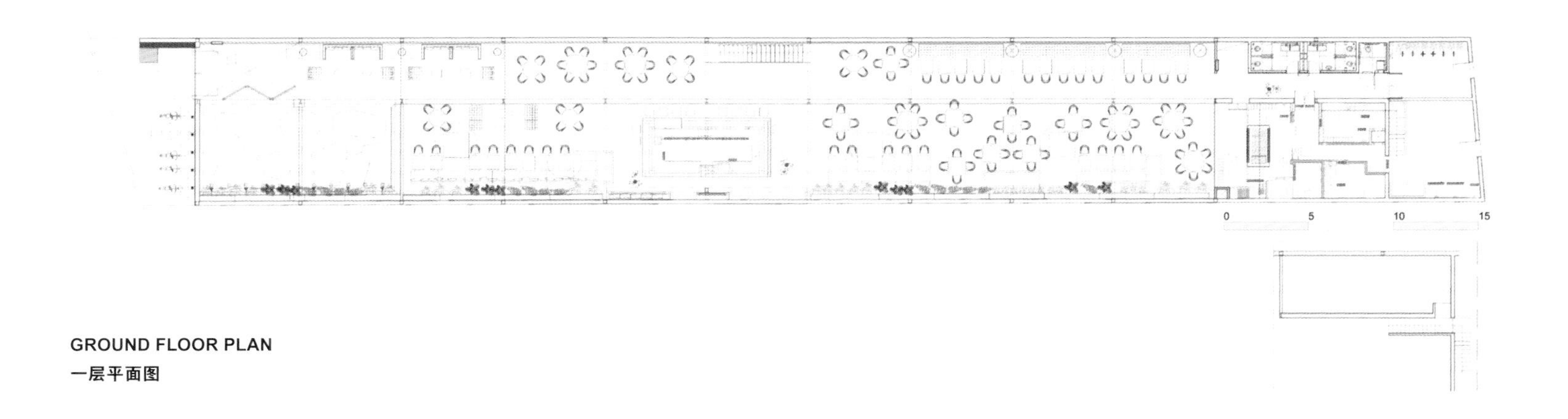

GROUND FLOOR PLAN
一层平面图

KKCG Offices Redesign

KKCG办公楼翻新

Completion date:
2012
Location:
Prague, Czech Republic
Designer:
VRTIŠKA • ŽÁK, RAW
Client:
KKCG
Photographer:
Kristina Hrabětová
Area:
8,500 sqm

竣工时间：
2012年
项目地点：
捷克，布拉格
翻新设计：
VRTIŠKA • ŽÁK设计工作室、RAW设计工作室
客户：
KKCG公司
摄影师：
克里斯蒂娜·哈贝托娃
面积：
8,500平方米

Project description:

Newly renovated office building in Prague was designed by Prague-based studio VRTIŠKA • ŽÁK, together with Brno-based RAW ATELIER. The client is an international business company, focused on an environmental sustainability. This project purely reflects natural approach and materials.

Dominant item of the building interiors is a 21-metre-high vertical garden, where the plants are grown. The moisture system is computer controlled. Another eye-catcher, once you enter the main lobby, is a white polygonal "interior façade" made out of lacquered aluminium profiles. It gives to the lobby a dynamic, yet calm feeling, all together combined with a lacquered glass cladding, Corian reception and a Barrisol light-up ceiling.

In this ground floor, you can find an office restaurant with a foyer façade-looking ceiling. Here the focus was placed in a horizontal position to make a floating effect of the ceiling and to cover the installations. The space is divided again with vertical green wall, with Lacobel cladding, wooden furniture, bar and a blackboard to communicate seasonal offers.

The meeting rooms in upper floors are made as a wooden box frame, implemented into an existing space. Meeting rooms are equipped with Privalight glass technology, which allows users to mild the glass only by touching a button.

Corridors are covered with vinyl woven carpets and equipped with custom-made doors. Office doors have a coloured glass on a side to give the space a fresh look and to be an orientation point in between business divisions as well.

Executive offices are made in Pandomo material, in combination with steel sheets claddings. These claddings can be used as a magnetic board to hang photographs or personal pictures, all together again in a combination

with wooden furniture.

Clubroom is a room equipped with hidden storage and a fireplace. The room zones are divided with a pile of wooden logs, stored in a metal frame.

Together with the spaces, also terraces were redesigned in a way to create more user-friendly surroundings and more natural touch.

In the building, there is also newly made employee fitness, created as a rough, yet cosy space with an accent on colours and materials. One wall mirror with built-in doors gives more space and more light to the interior. Toilets, cabins and showers are in a wooden block. The floor made of green rubber gives to the space the feel of energy and is well combined with red colour in the showers. The rest of the space is made in concrete structures and all the ceiling installations are exposed to create more industrial surroundings.

Together with the green floor in fitness, you can find a similar colour in the garage. The garage is made all in one green colour with the grass upper line element to reach a postive vibe. Very dominant is a white orientation system, placed on the floor and walls so you always know where to go or where are you right now.

布拉格的VRTIŠKA • ŽÁK设计工作室和布尔诺的RAW设计工作室共同负责KKCG公司办公楼的翻新。KKCG公司是一家国际企业，十分关注环境的可持续发展。在此次的办公楼翻新中，不论是设计的手法还是使用的材料，都体现了绿色环保的理念。

大楼室内最引人注目的要数21米高的垂直花园了。植物生长十分茂盛，湿度由电脑控制。另一大看点是由白色喷漆铝材制成的“栅栏墙”，一走进大厅就能看到。这一独特的设计让大厅显得极具动感，同时又营造出宁静的氛围。此外，大厅内还有漆面玻璃、可丽耐材质的接待台以及装有照明设备的法国巴力天花。

一楼还有员工餐厅，其中天花板的设计是亮点。设计师侧重水平方向的流线型设计，既让天花体现出流畅的动感，又巧妙地遮盖了天花内的各种设备。餐厅也进行了绿化，用一面绿墙将空间一分为二。Lacobel漆板玻璃、木质家具、吧台和黑板，营造出餐厅温馨宜人的环境。

楼上的会议室以木质框架结构为特色。会议室采用Privalight玻璃照明技术，只要轻触按钮，就能改变玻璃的照明强度。

走廊采用乙烯基纺织地毯，门都是专门定制的。办公室的门上有彩色玻璃，既装饰了空间，同时也在不同的办公部门之间起到指引的作用。

行政办公室地面采用磐多磨材料（Pandomo），墙面采用钢板，可以用作磁力板，贴一些照片在上面。行政办公室也跟餐厅一样，采用木质家具。

游乐室里有壁炉和隐藏的储物柜。设计师采用一摞圆木将空间进行分割，圆木安装在金属框架内。

除了这些空间以外，露台也进行了重新设计，目标是让空间使用起来更方便宜人，让环境更自然。

员工健身房是全新打造的一个空间。设计师侧重色彩和材料，空间看上去粗犷，不拘小节，却十分舒适。镜面墙上的门是嵌入式的，镜子让空间看上去更加宽敞、明亮。卫生间、储物间和淋浴间都设置在木质框架结构中。健身房地面采用绿色橡胶材质，让空间充满活力，并且跟淋浴间内的红色相辅相成。其余空间采用混凝土结构，天花板上的装置都裸露出来，营造出一种工业环境的气息。

跟健身房里的绿色地面相似，车库里采用了统一的绿色，也呈现出工业风格。车库里最引人注目的是导视的设计，墙面和地面上都有，随时告诉你现在在哪或者该往哪走。

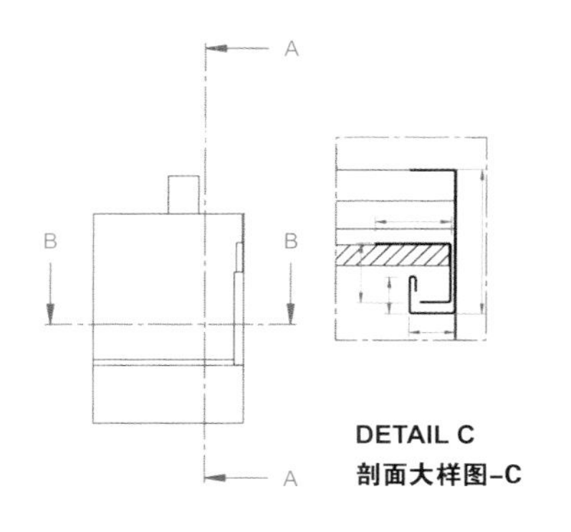

DETAIL C
剖面大样图-C

CROSS SECTION A-A
1. UA Profile
2. Aqua panel 15mm

横截面图-AA
1. UA钢材
2. 15毫米安耐板（Aquapanel）

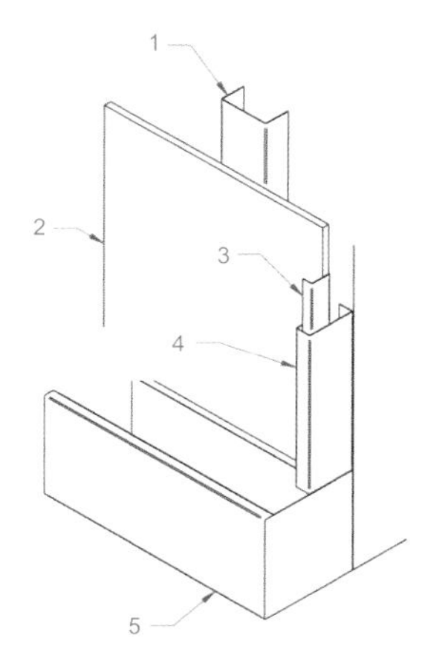

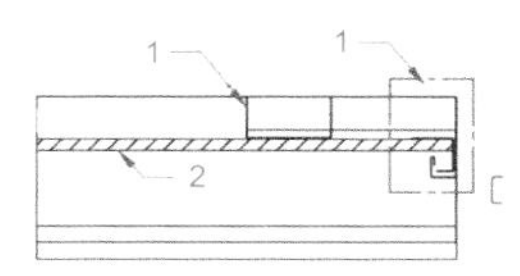

CROSS SECTION B-B
1. UA Profile
2. Aqua panel 15mm

横截面图-BB
1. UA钢材
2. 15毫米安耐板（Aquapanel）

1. UA profile
2. Aqua panel 15mm
3. Stainless steel inner cover
4. Stainless steel outer cover
5. Stainless steel container

1. UA钢材
2. 15毫米安耐板（Aquapanel）
3. 不锈钢内罩
4. 不锈钢外罩
5. 不锈钢容器

Aglaonema 粗肋草	Dracena 龙血树	Dracena surculosa 油点木	Dračinec vonný 巴西铁树	Epipremnum – šplhavník 绿萝
Fikus lyrata 琴叶榕	Maranta 竹芋	Monstera 龟背竹	Nolina 酒瓶兰	Philodendron cobra “眼镜蛇”蔓绿绒
Philodendron imperial green “绿帝王”蔓绿绒	Philodendron imperial red “红帝王”蔓绿绒	Spathiphyllum 白鹤芋	Syngonium 合果芋	

Les Néréides Flagship Store

蕾娜海台北旗舰店

Completion date:
2013
Location:
Taipei, Taiwan, China
Architect:
Chun-ta Tsao
Client:
Tun-Group
Photographer:
Ivan Chuang
Area:
145 sqm
Plants:
Tuberous sword fern, Parlor palm

竣工时间：
2013年
项目地点：
中国，台湾，台北
建筑设计：
曹均达
客户：
惇聚国际股份有限公司
摄影师：
庄博钦
面积：
145平方米
植物：
肾蕨、袖珍椰子

Project description:

The design concept for the flagship store originates from both the brand and the floor layout. The spirit of Les Néréides is perfectly projected to consumers through the ambiance and story behind the products. The style of Les Néréides carries the romantic feelings of a young French maiden and the unadulterated beauty of nature – one of the design concepts stems directly from this style. The selection of materials is the key to the brand. The main mediums are wood and enamel with a Southern French flare; elements of metal and marble are added, and then each piece is finished off with the Les Néréides "natural" touch.

The second design concept, floor layout, is a major component of the flagship store. With a multitude of materials, the narrow display space proved to be a hurdle to conveying the full essence of Les Néréides; hence the designers utilised this special long and narrow space to create "purposive" and "free" traffic flow. Consumers can appreciate products while moving about freely to experience the beauty of Southern France. The interior design also echoes each of Les Néréides series: the green wall at the entrance exhibits pieces which drew inspiration from flora; inside, the more expensive classic pieces are set against a more mature brown lacquer and oak embellished with metal; and the N2 line (a more youthful collection) uses bright colours and curving lines as the backdrop for a playful, mirrored wall to make consumers feel pampered. Each story told while meandering through the sections generates a deep connection between person, space, and product.

The project, completed in May 2013, features a green wall spanning almost ten metres long. The large green wall was adopted in the

interior to best convey the essence of Les Néréides, symbolising that pieces of Les Néréides were grown out of plants by putting display cabinets among plants. The contrast between nature and commerce leaves a strong impression.

A timing irrigation system is adopted for the green wall to ensure that when the shop is closed, an appropriate amount of water would be given, while in open hours, artificial lighting is provided to create suitable living conditions for the plants. A "mix-and-match" pattern is proposed to avoid the conventional problem of disorder, and small-leaf plants are put in a mixed way, including Tuberous sword fern, Parlor palm, Ficus pumila, Peperomia obtusifolia, Asparagus fern and Epipremnum aureum. In this way, the wall got a romantic yet natural texture, achieving a layered effect.

FRONT ELEVATION

1. Waterproofing lacquer
2. Light box with non-opaque poster
3. Teakwood with brush finish
4. See detail drawing
5. Floor/oak board
6. Green wall
7. Air damper
8. Floor/rotor brick
9. Oil lacquer, Pantone PSM566
10. Rose gold titanised panel

立面配置图（前区）

1. 防水泥粉光漆处理
2. 灯箱/贴透光海报
3. 柚木钢刷面自然拼
4. 见详图
5. 地板/橡木木地板
6. 植物绿墙
7. 空气门
8. 地板/火头砖人字频
9. 油性喷漆 Pantone PMS566
10. 玫瑰金镀钛板

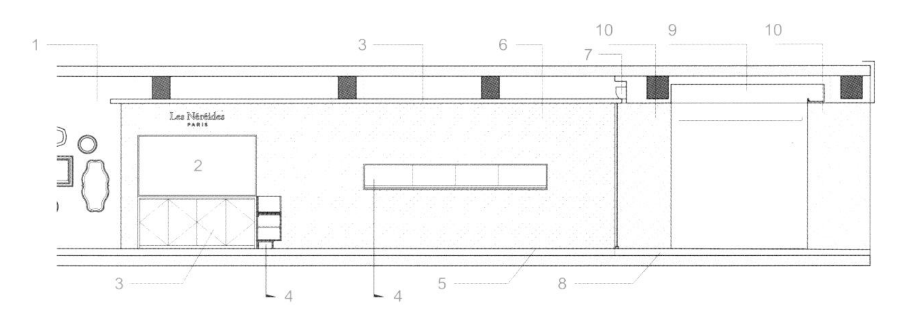

FRONT ELEVATION

1. Melamine panel
2. ICI latex paint
3. See detail drawing
4. Tattered lacquer
5. Oak borad
6. Waterproofing lacquer
7. Green wall

立面配置图（前区）

1. 美耐板
2. ICI乳胶漆
3. 见详图
4. 油漆仿旧处理
5. 橡木木地板
6. 防水泥粉光漆处理
7. 植物绿墙

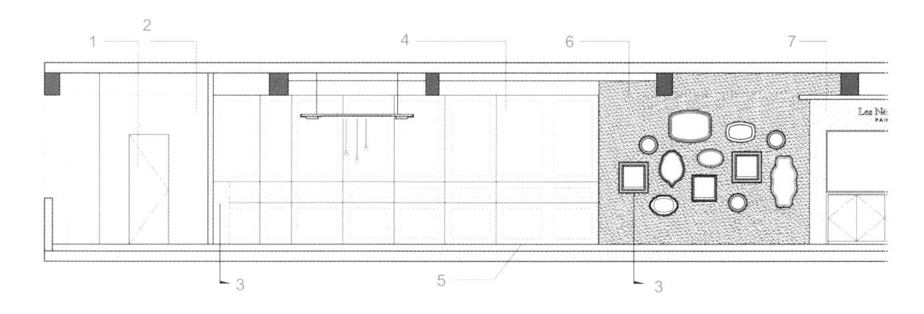

The interior irrigation system provides water from the top of the wall. Every planting box on the wall has draining holes, so with a fixed amount of water provided from the top, every box will be fed with enough water. Under the boxes at the bottom, a drainage channel is placed to drain spare water. Top irrigation is conducted once per day, lasting for about 15 minutes.

Different from exterior green walls, small-leaf plants that are suitable for indoor planting are adopted in the project, creating a more romantic ambiance.

蕾娜海台北旗舰店的设计理念来自两方面，一方面是品牌自身，另一方面是室内平面布局。设计师通过室内氛围的营造，将法国蕾娜海珠宝的品牌精髓及其珠宝饰品背后的故事完美地传达给顾客。蕾娜海品牌珠宝体现出一种法国少女特有的浪漫感觉以及大自然的纯粹的美——本案的设计理念之一就源于这种风格。材料的选择对于树立品牌形象至关重要。主要材料是木材和搪瓷，搪瓷有着法国南部搪瓷特有的光芒。此外，还增加了金属和大理石元素，都像每一件蕾娜海珠宝饰品那样打磨出自然的光泽。

设计理念的另一方面是平面布局，是这家旗舰店设计的重要组成部分。店内材料多种多样，展示空间很狭窄，这对于蕾娜海珠宝饰品的陈列和品牌精髓的传达是个问题。鉴于此，设计师利用既定的狭长空间打造了流畅的交通流线。顾客可以一边浏览饰品，一边在店内自由行走，体验法国南部特有的美感。室内设计用不同的手法体现蕾娜海品牌不同系列的珠宝：入口处的绿墙展示的饰品，其设计灵感来自花朵；店内，更高档的古典饰品的陈列以镶嵌金属装饰的棕色橡木板为背景，更显奢华；N2系列饰品针对年轻人，陈列背景采用鲜艳的色彩和弯曲的弧线——一面妙趣横生的镜面墙，让顾客体验购物的乐趣。在店内各个空间行走，你会体验到每个空间设计背后的构思，体验人、珠宝、空间三者融为一体的感受。

蕾娜海台北旗舰店于2013年5月完工。本店特色在于将近10米长的绿墙。基于蕾娜海的品牌精神及大自然的元素，所以使用大面积的绿墙，想要塑造饰品是从植物里头生长出来的，所以刻意将展示柜放置在植物中间，形成自然与商业化的对比，加强给人的印象。

绿墙使用定时灌溉系统，在闭店后能够定量给水，加上开店人工灯光的照射，使植物能够在室内存活生长。绿墙的立面设计以混搭为主，试图克服一般绿墙规律的不协调感，并选用小叶的植物作为混搭，如：肾蕨、袖珍椰子、雪茄、圆叶椒草、文竹、黄金葛等。如此一来在质感上就会比较浪漫且自然，区块上也使用像迷彩的配置，达到有层次且自然的效果。

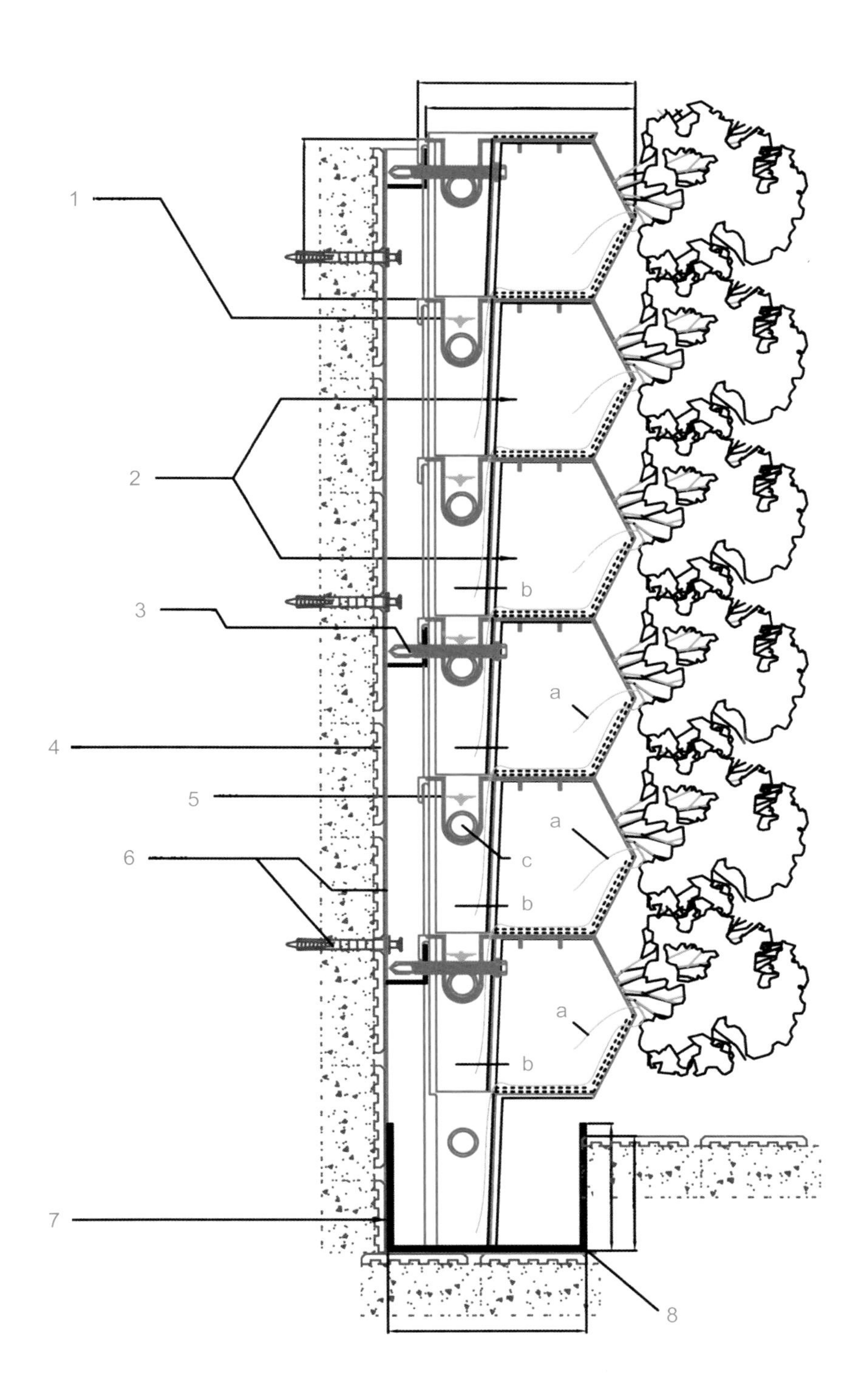

CONSTRUCTION DETAIL – PROFILE IN ELEVATION

1. Overflow level
2. Water retention planter outer slot (planting density: 64PS/m^2)
3. Galvanised angle steel & fixing self-tapping screw (with interval distance no more than 50cm)
4. Vertical exterior wall
5. Overflow level
6. Wall-mounted galvanized angle steel & expansion bolt
7. Stainless steel key trench
8. Pre-digged by client
   a.Growing medium: siphon cloth
   b.Water reserve
   c.Horizontal T-branch pipe

剖立面施工图示意图

1. 满水自然溢水水位
2. 保水蓄水植栽外槽（植栽密度64PS/m^2）
3. 镀锌角钢与固定补强自攻螺丝（间距皆小于50厘米）
4. 结构垂直外墙
5. 满水自然溢水水位
6. 墙面镀锌角钢与膨胀螺丝
7. 不锈钢截水槽
8. 业主先行地面挖除
   a. 植栽介质：虹吸布
   b. 蓄水区
   c. 左右横向三通管

FRONT ELEVATION
1. Teakwood with brush finish
2. Oil lacquer, Pantone PMS566
3. Tattered lacquer, Pantone PMS566
4. Wall paper
5. Waterproofing lacquer
6. Oil lacquer, Pantone PMS566
7. Floor/rotor brick
8. Rose gold titanised panel
9. See detail drawing
10. Floor/oak board
11. Mirror

立面配置图（前区）
1. 柚木钢刷面自然拼
2. 油性喷漆Pantone PMS566
3. 油漆仿旧处理Pantone PMS566
4. 面贴壁纸
5. 防水泥粉光处理漆
6. 油漆仿旧处理Pantone PMS5483
7. 地板/火头砖
8. 玫瑰金镀钛板
9. 见详图
10. 地板/橡木地板
11. 明镜

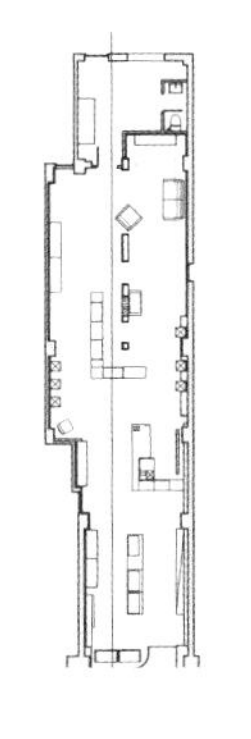

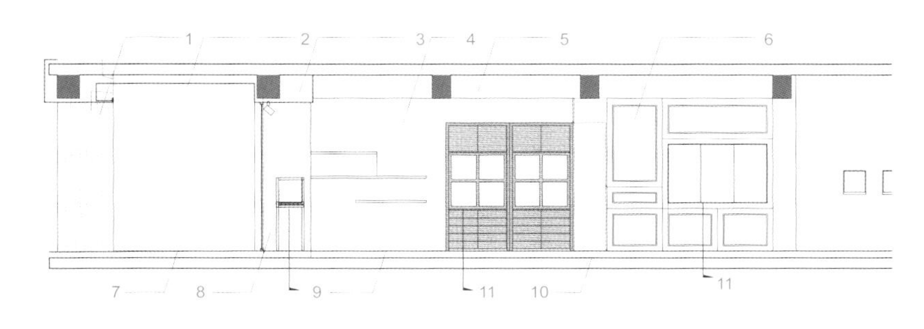

REAR ELEVATION
1. Wall paper
2. Dyed oak wood
3. Tattered lacquer, Pantone PMS566
4. ICI latex paint
5. See detail drawing

立面配置图（后区）
1. 面贴壁纸
2. 橡木染色处理
3. 油漆仿旧处理Pantone PMS5483
4. ICI乳胶漆
5. 见详图

室内的排水系统，将给水都设置在最上头。因为每个栽培盒都有设计排水孔，所以在定量的供水下，每个栽培盒可以吸收到足够的水分，然后往下流，直到最后一个栽培盒，在最后一个栽培盒下设计排水沟，将多余的水排出。每日供水一次，一次约15分钟。

与一般外立面绿墙不同的是，本案采用更多小叶的植物，更符合室内栽培的植物，感受上也比较浪漫。

Main Street Green Wall

罗维戈主街室内绿墙

Location:
Rovigo, Italy
Architect:
Studio Paparella
Photographer:
Ludovico Guglielmo
Area:
54 sqm

项目地点：
意大利，罗维戈
建筑设计：
帕帕雷拉建筑设计工作室
摄影师：
卢多维科・古列尔莫
面积：
54平方米

Project description:

The green wall is composed by Grüne Wand® system. It is formed by stainless steel framework with modular system of 40×60cm and sheets of vegetation.

Plates, in turn, are composed by three layers:
* at the bottom, polystyrene panel;
* in the middle, phenolic foam;
* an out layer of vegetation.

The vegetation elements are cultivated at a farming company for about 14-18 weeks before installation. The system pledges:
* the advantages of plantation on synthetic material;
* control of humidify activity of the air because 90% of the evaporation happens through substratum.

The green wall is provided with tank bedded in a technical compartment and, therefore, technical components are barred from the public area.

When the interior space is provided with the green wall, the inside microclimate is evaluated by a specialist who makes survey in order to decide: the size of the partition and the regulation of the light conditions (lighting system around the wall must be regulated with particular enlightening elements).

Complete setup is provided with water supply system and automatic decentralised control unit of both irrigation and fertilisation. The monitoring is managed by decentralised computers and connected by GLT internet systems with alarm. Through modem the control system checks devices, hydraulic pumps and technical parameters.

Supported by digital compensator in the control cabinet, the water supply acts as a function of ambient air humidity and time. The modern technical assistance can modify all the parameters.

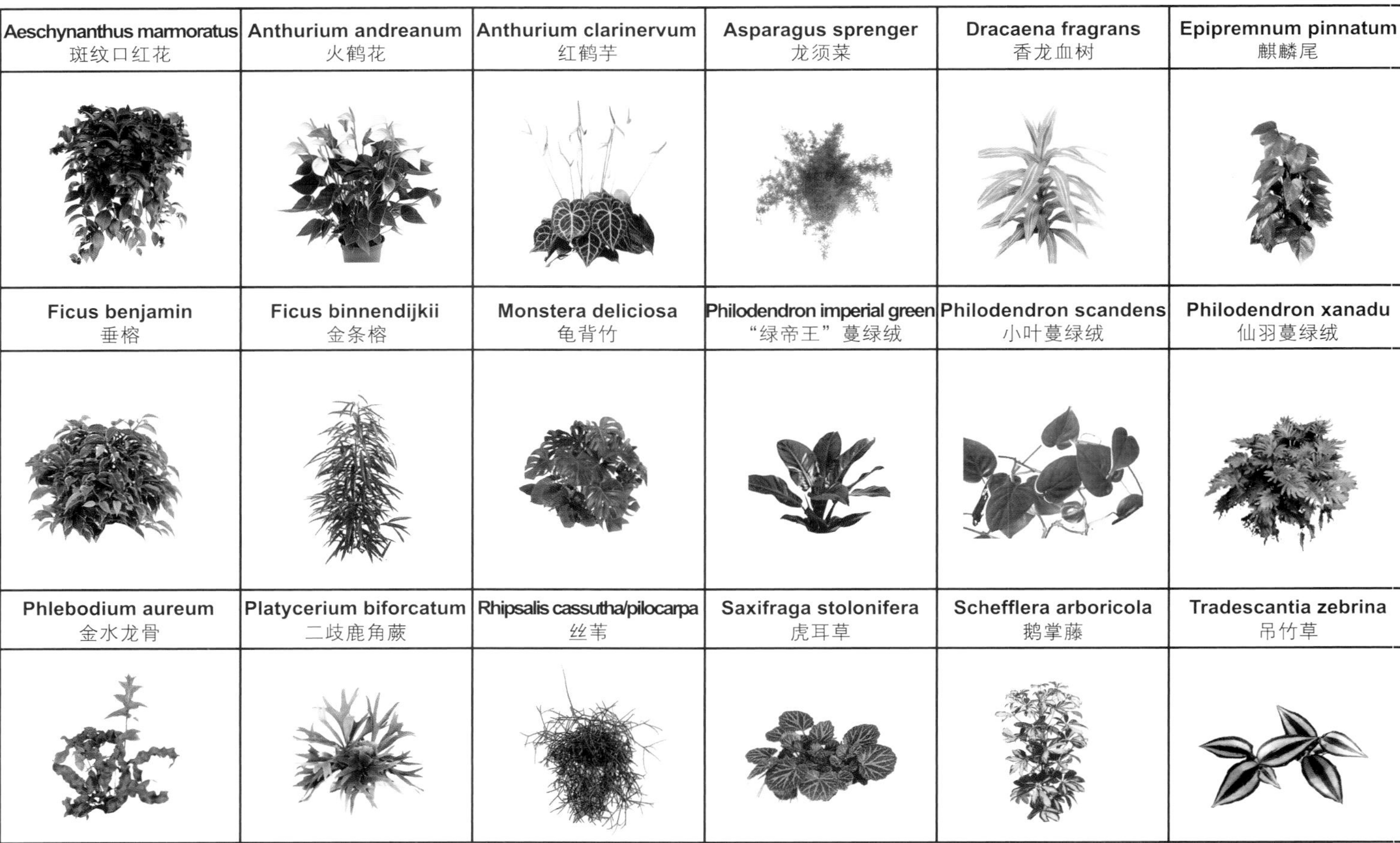

Aeschynanthus marmoratus 斑纹口红花	Anthurium andreanum 火鹤花	Anthurium clarinervum 红鹤芋	Asparagus sprenger 龙须菜	Dracaena fragrans 香龙血树	Epipremnum pinnatum 麒麟尾
Ficus benjamin 垂榕	Ficus binnendijkii 金条榕	Monstera deliciosa 龟背竹	Philodendron imperial green “绿帝王”蔓绿绒	Philodendron scandens 小叶蔓绿绒	Philodendron xanadu 仙羽蔓绿绒
Phlebodium aureum 金水龙骨	Platycerium biforcatum 二歧鹿角蕨	Rhipsalis cassutha/pilocarpa 丝苇	Saxifraga stolonifera 虎耳草	Schefflera arboricola 鹅掌藤	Tradescantia zebrina 吊竹草

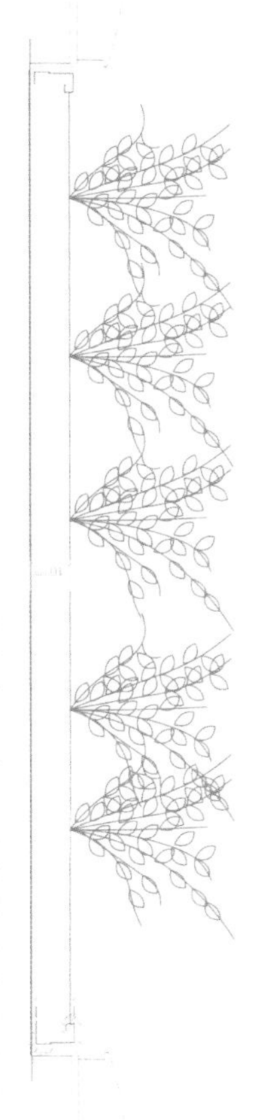

DETAIL 细节图

这面绿墙采用德国格伦·万德公司开发的绿墙设备，以不锈钢框架结构为基础，框架内的模块规格为40厘米×60厘米，带有覆盖着植被的板材。植被板材由三个层次构成：
- 最内层是聚苯乙烯板材
- 中间层是酚醛泡沫塑料
- 最外层是植被

植物先在一家农业公司培育14～18周，然后栽种到绿墙上。这面绿墙有以下技术特色：
- 植物生长于合成材料上
- 能控制空气湿度（因为90%的蒸发作用都通过下层的培养基发生）

水箱隐藏在一个隔间里，从外部看不到绿墙的技术构件。室内安装了绿墙之后，专家专门对室内微气候进行了评估，以便决定这面绿墙的体量是否合适，以及光照条件应该如何调节（绿墙周围必须有专门的照明设备，调节植物的光照条件）。

最后是供水设计，以及分散控制式灌溉和施肥系统。用分散式电脑控制系统实施监控，并与警报系统相连。通过解调器，控制系统能够检查设备、液压泵和技术参数。通过控制箱内的数控变压器，供水系统能够控制空气湿度和供水时间。现代技术的使用能够修改所有相关参数。

Papadakis Integrated Sciences Building, Drexel University

德雷塞尔大学帕帕扎基斯综合科学楼

Completion date:
2011
Location:
Philadelphia, Pennsylvania, USA
Architect:
Diamond Schmitt Architects
Photographer:
Tom Arban
Area:
13,935 sqm

竣工时间：
2011年
项目地点：
美国，宾夕法尼亚州，费城
建筑设计：
戴蒙德&施密特建筑事务所
摄影师：
汤姆·阿本
面积：
13,935平方米

Project description:

Diamond Schmitt Architects of Toronto designed the Papadakis Integrated Sciences Building (2011) at Drexel University in Philadelphia to be a highly sustainable building and to set a new standard for green building design at a United States university. The building received LEED Gold certification in 2012.

The 13,935-square-metre facility houses 44 research and teaching laboratories for biology, organic chemistry and biomedical engineering. A ground floor 250-seat auditorium and café are adjacent to a large atrium with a dramatic coiled staircase and a 23-metre-tall biofilter living wall – the largest installation of its kind to date in the U.S.

The building creates a living room, a gathering space for the Drexel community which didn't exist before, and the design intent is to allow students and faculty to come together to work and socialise in the atrium, in labs filled with natural light, and in lounges or "collaboratories" located in a three-storey transparent tower.

The biofilter living wall brings a natural ecosystem indoors that purifies the air, reduces energy costs for heating, cooling and humidifying a building and creates a healthier, more pleasing indoor environment. Anecdotal evidence points to a decrease in sick days and higher productivity in buildings with biofilter living walls.

Diamond Schmitt Architects pioneered the commercial application of the biofilter working with its developer, Nedlaw Living Wall Inc., at the University of Guelph Humber

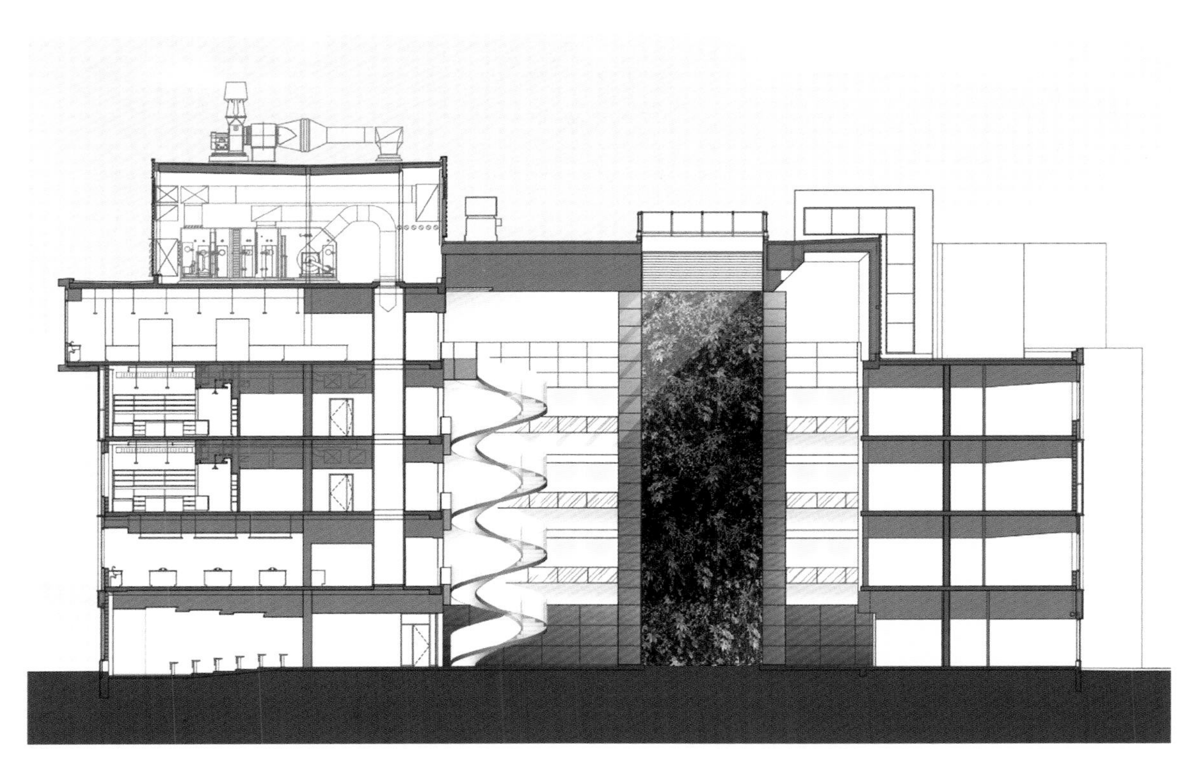

SECTION　剖面图

campus in Toronto in 2004. Since then, the two firms have designed and installed more than a dozen living walls, including Cambridge City Hall, the Royal Botanical Gardens, the University of Windsor, Corus Quay office building and Centennial College in Toronto and at the University of Ottawa.

The biofilter living wall works by embedding plants in a synthetic rooting media. These tropical plants are nourished hydroponically by water that is pumped to the top of the wall and flows down. Contaminated room air is drawn through the beneficial microorganisms growing in the root zone of the plants that utilise airborne pollutants as food and break them down into their benign components, water and carbon dioxide. These living wall biofilters eliminate 80 percent of airborne chemicals emitted from building materials and furnishings, combustion pollutants like carbon monoxide and toxic particles and biological contaminants such as moulds and bacteria.

Rather than draw air from the outside, which has to be heated or cooled depending on the season, recycled biofilter air reduces the need and cost for modifying outdoor air. Most living walls are fully integrated into a building's air handling system. They draw air through the biofilter and send the clean air into the building through the supply airshafts.

Biofilters big and small are proving effective in reducing the presence of volatile organic compounds in the air. They are best situated in areas that have plenty of natural light to assist in the growth of the plants. At five storeys high, the Drexel University biofilter is the largest to date on a U.S. campus. Diamond Schmitt Architects designed a larger, six-storey wall as part of the Faculty of Social Sciences at the University of Ottawa, which opened in late 2012.

多伦多的戴蒙德&施密特建筑事务所负责美国费城德雷塞尔大学的帕帕扎基斯综合科学楼的室内绿化设计。这栋可持续的绿色建筑为美国大学的绿色建筑设计树立了新的标杆。2012年，这栋大楼获得了美国绿色建筑委员会的LEED金级认证。

这栋大楼建筑面积为13,935平方米，内有44个科研与教学实验室（学科有生物学、有机化学和生物医学工程）。一楼中庭旁边是礼堂（能容纳250人）和咖啡厅。中庭非常宽敞，螺旋形楼梯尤其吸引眼球。约23米高的绿墙是一面活的“生物过滤墙”——这是美国当时同类绿墙中面积最大的。

绿墙的设计让德雷塞尔大学的师生关系更紧密了，甚至可以说建立了“德雷塞尔社区”。中庭成为师生聚会和活动的场所。此外，洒满阳光的实验室和三层高的透明塔楼中的休闲区也都是师生交流的好去处。

通过这面“生物过滤墙”，楼内建立了自然生态系统，不仅能净化空气，降低建筑供热、制冷、加湿等能耗，而且创造了更加健康、更加宜人的室内环境。已有证据表明，有室内绿墙的建筑在较好的天气里会创造更好的室内气候。

早在2004年，戴蒙德&施密特建筑事务所就与绿墙设备开发商Nedlaw公司合作，在加拿大圭尔夫大学汉博分校探索绿墙的实际应用。从那时起，这两家公司就联手设计并实践了10多个绿墙工程，包括剑桥市政厅、皇家植物园、温莎大学、鲁斯娱乐公司办公楼、多伦多百年理工学院和渥太华大学等。

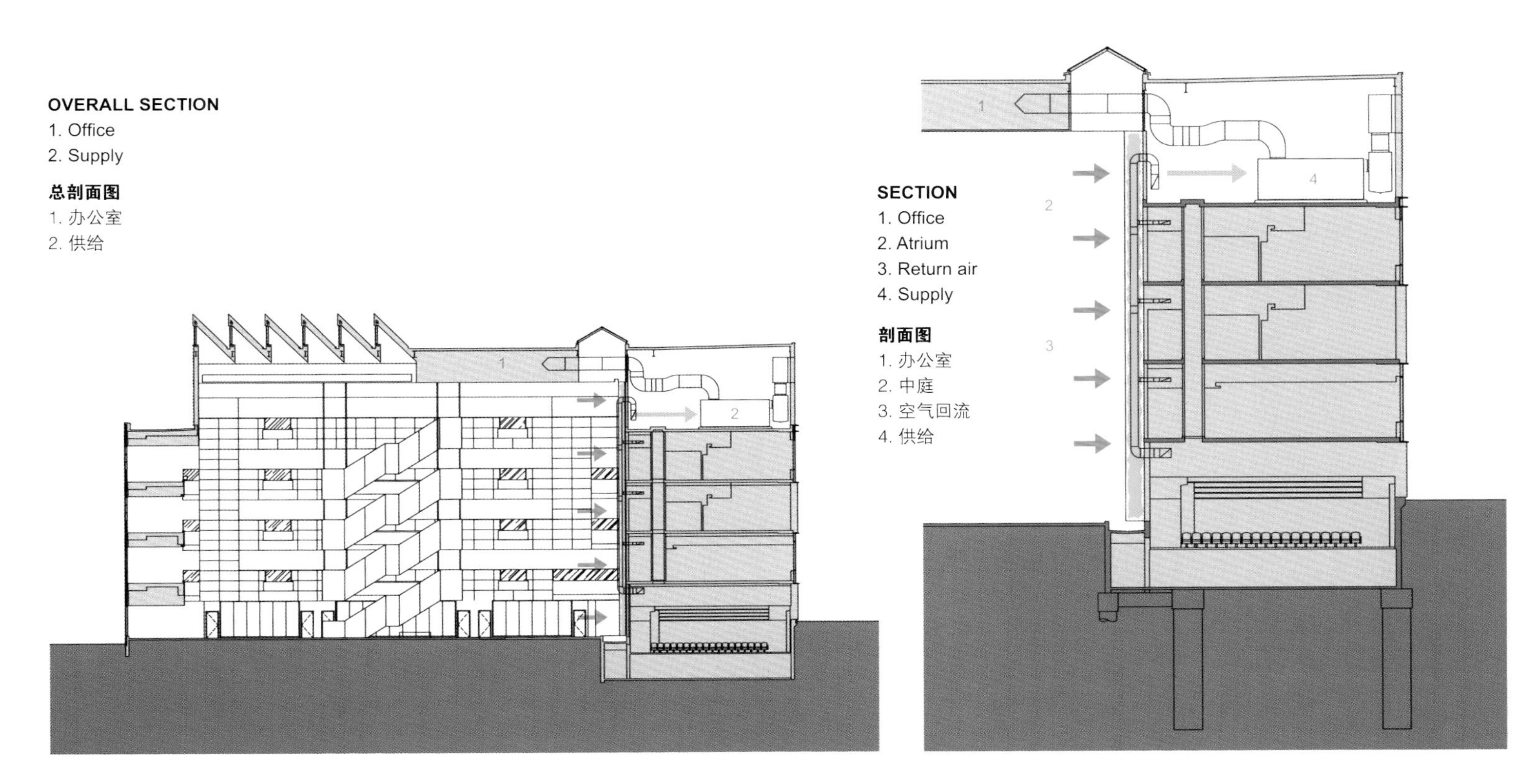

OVERALL SECTION
1. Office
2. Supply

总剖面图
1. 办公室
2. 供给

SECTION
1. Office
2. Atrium
3. Return air
4. Supply

剖面图
1. 办公室
2. 中庭
3. 空气回流
4. 供给

生物过滤墙将植物种植在合成的根系生长介质中。种植的都是热带植物，用营养液培育，营养液用水泵输送到绿墙顶端然后流下来。植物根系中滋生的微生物会净化受到污染的室内空气，植物把空气中的污染物作为养分，将其分解成良性的组成成分、水和二氧化碳。这种生物过滤墙能消除建筑和装修材料中释放的80%的空气化学物质、燃烧污染物（如一氧化碳）以及有毒微粒和生物污染物（如霉菌和细菌）。

如果从室外引进气流的话，需要根据季节进行供暖或降温，生物过滤墙却不涉及这个问题。植物能够对室内空气进行更新，降低了改善户外空气的需求和成本。大多数绿墙都能完全融入建筑的空气处理系统。绿墙通过生物过滤作用，利用通风井将新鲜空气输送到大楼各个角落。

生物过滤墙不论大小，都已经证明对减少空气中的挥发性有机化合物有显著效果。生物过滤墙最好设置在有充足的自然光线的地方，以便保证植物的茂盛生长。德雷塞尔大学的生物过滤墙有5层楼高，是当时美国大学中最大规模的绿墙。戴蒙德&施密特建筑事务所随后又为渥太华大学社会科学系大楼设计了一面更大的、6层高的绿墙，于2012年底竣工。

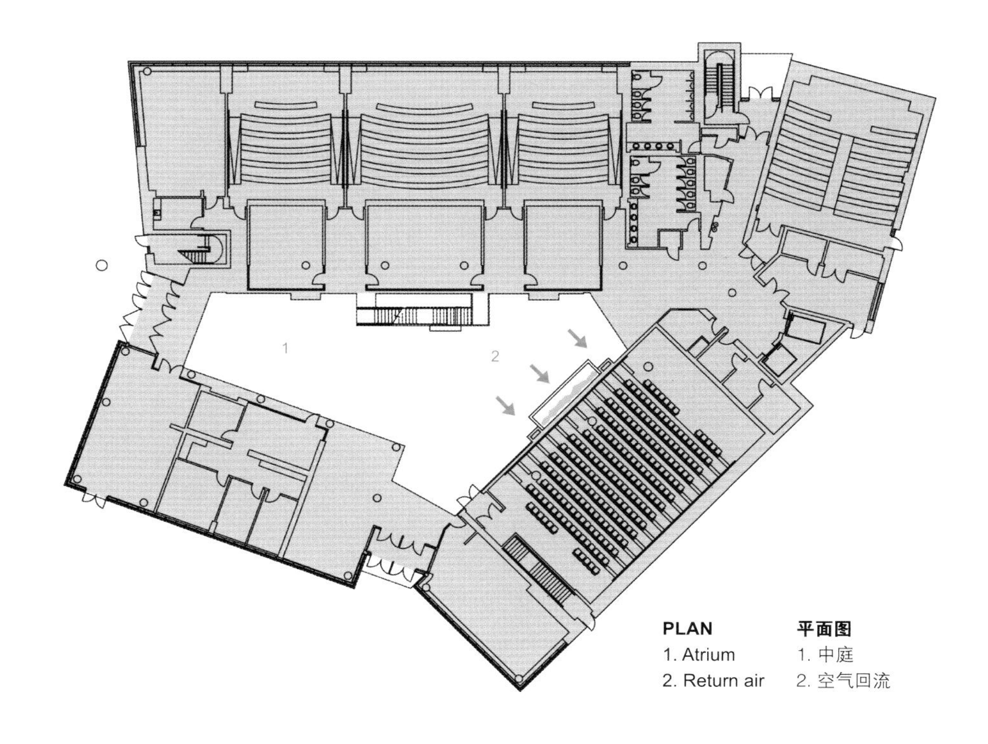

PLAN
1. Atrium
2. Return air

平面图
1. 中庭
2. 空气回流

Stücki Shopping Centre

斯塔基购物中心

Completion date:
2012
Location:
Basel, Switzerland
Architect:
CREAplant AG, LignaPlan AG
Area:
770 cm x 380 cm

竣工时间：
2012年
项目地点：
瑞士，巴塞尔
建筑设计：
瑞士CREAplant公司、瑞士LignaPlan公司
面积：
770厘米 x 380厘米

Project description:

Walls are transformed into green artworks.

The Wonderwall offers infinite possibilities of design. Peerless textures of an enormous variety of diverse growing types of vegetation allow a vertical arrangement ranging from stringent geometrical schemes of greenery to organic formed patterns. Vertical spaces are transformed into animate walls: stylish, individual and amazingly energetic.

· Increased humidity: Measureable accumulation up to the ideal value of 40 to 50% humidity.
· Improved air quality: The wall of plants reduces pollutants and provides oxygen.
· Enhanced sense of well-being: The natural green surface enhances the subjective well-being and alleviates stress.
· Optimised acoustic: The construction of Wonderwall has a sound-absorbing impact.
· Three-dimensional design: Selected types of vegetation grow not only upwards but also horizontally out of the wall.
· Flexible application: Wonderwall can be detached or attached to all sorts of rear wall.
· Ideal space utilisation: Induced to the vertical arrangement, no deprivation of valuable office or sales area occurs.

The vertical garden provides an improved indoor climate.

The vertical garden Wonderwall is an innovative, aesthetically appealing, yet green solution improving the indoor climate. The allover greenery has several positive characteristics:

Wonderwall: The system
A Wonderwall can be detached or attached to all sorts of rear wall. A vegetation mat composed of several geotextile tiles is fixed to a subconstruction of stainless steel. Each layer of the prefabricated mat has its separate function. The particularly selected greenery will be placed into pockets along the outside

NEWS

vegetation mat. An integrated irrigation system supplies the planting with moisture as well as fertiliser demand-actuated and automatically regulated. The nutritional liquid irrigates through the vegetation mat based and dispenses equally onto the entire surface. The idle water streams into the drainage system thence into a water reservoir creating a closed circuit without any loss of water.

The interior landscapers CREAplant

The Wonderwall constitutes the highlight of this attractive exhibition, a 20-square-metre large vertical garden consisting of more than 100 exotic plants. Wonderwall is CREAplant's most impressive and groundbreaking speciality.

将墙面变为绿色艺术

“奇幻绿墙”提供了无限的设计可能性。各色品种的植被令人眼花缭乱，构成独特的绿色质感。并且，垂直绿化技术让绿墙上的植被造型随意变换，从严格的几何图案到灵活的有机形状。单调的墙面变身为生机勃勃的墙上花园，既时尚，又新颖，而且让空间充满活力。

- 增加湿度：可以达到理想湿度（40%～50%）
- 改善空气质量：绿墙能够减少污染物，释放氧气
- 有助于身心健康：自然的绿色植物有助于人们的身心健康，使人心情愉悦，减少压力
- 吸音降噪：绿墙有吸收噪声的功效
- 3D设计：精心选择的植物品种不仅向上生长，也会横向蔓延，伸出墙外
- 灵活运用：“奇幻绿墙”可以安装在各种墙面上
- 理想的空间利用：垂直绿化不占用珍贵的办公空间或销售空间

垂直花园改善室内气候
垂直花园是改善室内气候的一种既美观又环保的创新手段。绿墙的设计有以下特征：

“奇幻绿墙”——完备的绿墙系统
绿墙可以安装在各种墙面上。植被层由多层土工织物构成，安装在不锈钢底座上。每一层都有不同的功能。专门选择的植物种植在最外一层的凹槽里。一体化的灌溉系统为植物提供水分和肥料，而且能够按需供应，自动控制。营养液通过下方的织物层进行供给，均匀分配到整面绿墙上。多余的水流入排水管道，再进入储水池，形成一个循环回水系统，不会造成水的浪费。

CREAplant室内景观设计公司
这面绿墙面积20平方米，包含100多种外来植物，是购物中心内的视觉焦点。“奇幻绿墙”是CREAplant公司推出的最具创意、最令人难忘的绿化设计。

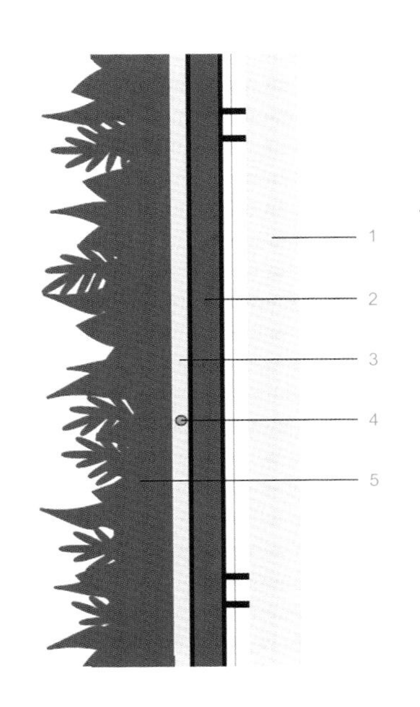

DETAIL
1. Rear wall on site
2. Subconstruction
3. Vegetation sheets
4. Irrigation system
5. Plants

细部大样图
1. 原有的后墙
2. 加建结构
3. 植被层
4. 灌溉系统
5. 植物

Anthropologie

"人类学"女装店

Location:
Huntsville, Alabama, USA
Architect:
EOA/Elmslie Osler Architects
Photographer:
Robert Reck
Area:
930 sqm

项目地点：
美国，阿拉巴马州，亨茨维尔
建筑设计：
EOA建筑事务所
摄影师：
罗伯特·雷克
面积：
930平方米

Project description:

This is an inspiring place to shop for women's clothing, accessories and home decor. Anthropologie offers a one-of-a-kind and compelling shopping experience that makes women feel beautiful, hopeful and connected.

A lush, green, vertical landscape was envisioned as a new kind of façade for Anthropologie's presence in Huntsville, Alabama. EOA/Elmslie Osler Architect's vision for the shop turned into what was at the time the largest Green Wall installation in North America. The soil-based walls are constructed of 2'x2'x3" panels of a variety of sedum genus. The living walls will bloom in the spring and stay green in the colder months when they will provide extra insulation to reduce energy use. In the summer, the south- and southeast-facing walls will absorb UV rays, cooling the interior of the building. The walls are bordered by white washed slats of wood screens and were conceived as an intuitive response to the fences that wrap the rich landscape of the south. As the living walls change through the seasons the plant life will bring a natural textural element into the anonymity of the typical suburban lifestyle centre. Ultimately the walls help to support an awareness of the environment and establish a more modern way of looking at nature. EOA designed several shops for the client, each with the challenge of creating a distinct response to factors such as the landscape, culture and history of the location while maintaining the core value of their very successful brand.

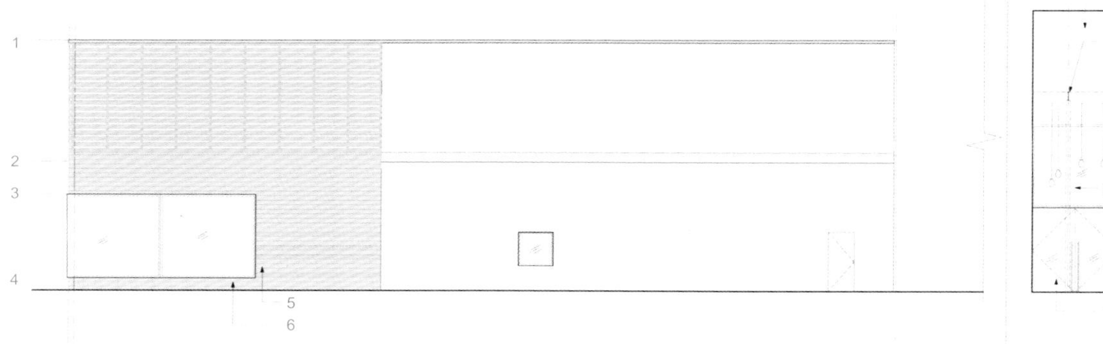

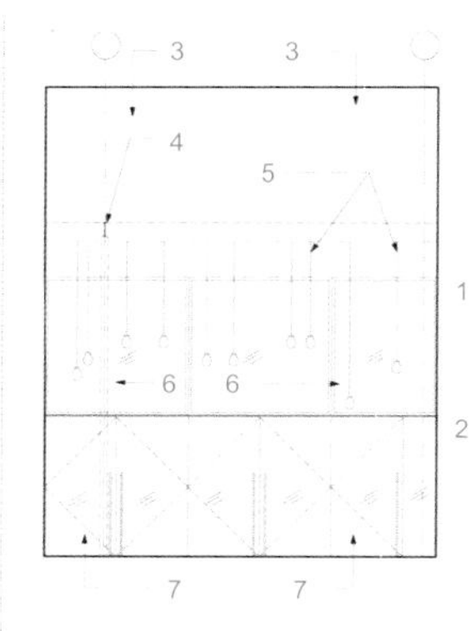

NORTH ELEVATION
1. Top of louvres
2. Top of precast panel
3. Top of opening
4. Top of sill
5. Floating metal letter sign
6. Reclaimed barn wood horizontal siding

北侧立面图
1. 遮光栅格顶部
2. 预制板材顶部
3. 开窗顶部
4. 窗台顶部
5. 悬浮的金属字招牌
6. 回收利用的水平方向护墙板

ENTRY ELEVATION
1. Top of transom
2. Top of entry
3. Existing cement plaster
4. Exposed structure
5. Pendant exterior light fixtures
6. Aluminium storefront transom
7. Frameless "herculite" glass entry doors horizontal siding

入口立面图
1. 气窗顶部
2. 入口顶部
3. 原有的水泥抹面
4. 裸露的结构
5. 室外悬垂灯具
6. 铝制店面气窗
7. 无框钢化玻璃门

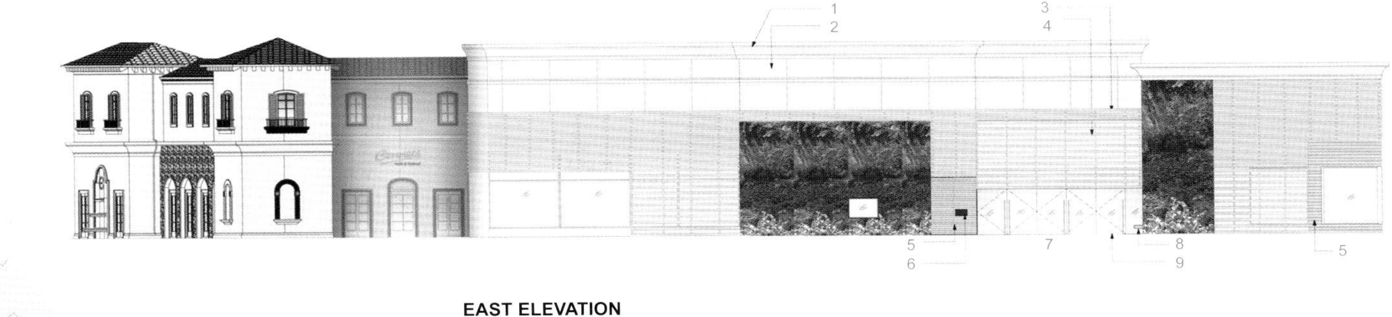

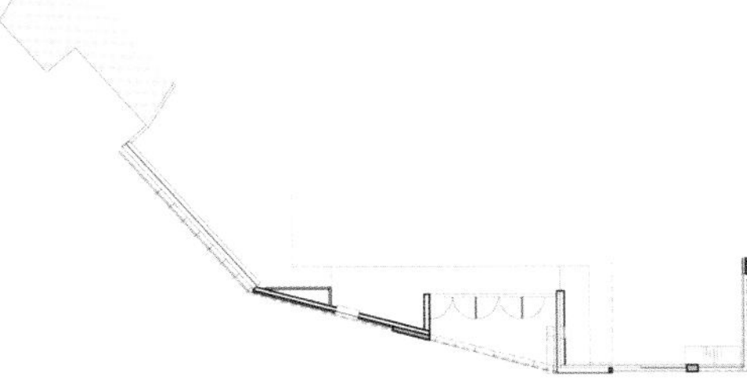

EAST ELEVATION

1. Cornice at parapet
2. Plaster stucco BFS
3. Reclaimed cypress wood horizontal boards w /2-1/2" spacing (O/EIFS)
4. Reclaimed cypress wood horizontal boards w/6" spacing (O/Glazing)
5. Reclaimed cypress wood horizontal siding, stacked
6. Laser cut backlit metal sign
7. Recessed entry
8. Reclaimed word bench
9. Recessed entry beyond

东侧立面图

1. 护墙檐口
2. 灰泥
3. 回收利用的柏木木板（水平方向拼贴，间隔6.35厘米，后面是外墙保温层）
4. 回收利用的柏木木板（水平方向拼贴，间隔15.24厘米，后面是玻璃幕墙）
5. 回收利用的水平方向柏木护墙板
6. 金属店招（激光切割，背光照明）
7. 缩进式入口
8. 回收利用的木质长椅
9. 缩进式入口内部

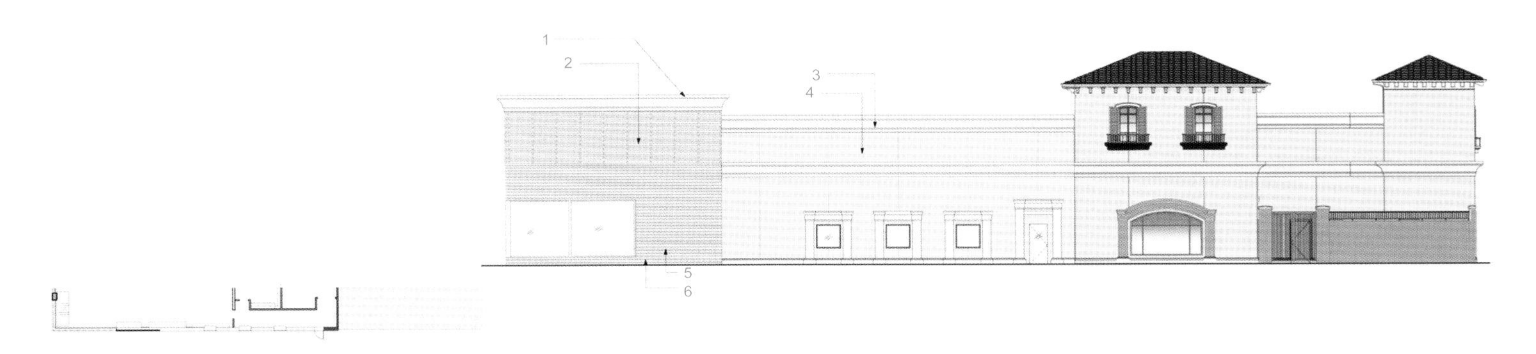

NORTH ELEVATION

1. Metal coping
2. Reclaimed cypress wood horizontal boards w/2-1/2" spacing (O/EIFS)
3. Cornice O/Parapet
4. Color of plaster stucco EIFS to match adjacent
5. Floating metal letter sign
6. Reclaimed cypress wood horizontal siding

北侧立面图

1. 金属压顶
2. 回收利用的柏木木板（水平方向拼贴，间隔6.35厘米，后面是外墙保温层）
3. 护墙檐口
4. 外墙保温层的灰泥颜色与旁边相匹配
5. 悬浮的金属字招牌
6. 回收利用的水平方向柏木护墙板

Anthropologie was founded in 1992. Sixteen years after being founded in Wayne, Pennsylvania, this project was completed in January 2008. The green wall brings textural modernity to the elevation of this typical suburban lifestyle centre. Women's fashion retailer Anthropologie remains a destination for women wanting a creative mix of clothing. It is a unique world of beautiful fashion.

这是美国阿拉巴马州亨茨维尔市的一家经营女性服装、配饰以及家居装饰品的店铺。外墙的垂直绿化让这家店铺别具一格，带来超凡的购物体验，让光顾这里的女顾客感觉自己更美丽、更自信、更健康。

长满茂盛植被的外墙代表了这家店铺崭新的形象。EOA建筑事务所打造的这几面绿墙是当时北美面积最大的绿墙。墙面上安装了种植槽，里面有土壤，种植了各种景天属植物。绿墙上的植物春季生长最为茂盛，之后较为寒冷的几个月里还能保持绿色的状态，对建筑起到保温的作用，降低了建筑能耗。夏季，朝向南侧和东南这两个方向的绿墙会吸收紫外线，对室内起到降温的作用。南侧外立面上白色木栅栏的设计非常巧妙，跟绿墙在一起，好像是草坪围栏。绿墙上的植物随着季节生长变化，赋予这栋建筑以自然的质感，为市郊一成不变的形象增添了一抹灵动的元素。此外，绿墙有助于提升人们的环境意识，让人们对自然树立起更现代的看法。EOA建筑事务所为这家客户设计了好几个门店，每一个都因地制宜，侧重不同的方面，有景观、文化、当地历史等，核心宗旨只有一个——树立品牌价值。

“人类学”这个品牌（Anthropologie）始创于1992年，创始地在美国宾夕法尼亚州的韦恩。本案于2008年1月竣工，距其初创已有16年的历史了。现代感十足的绿墙形象让这家店铺在这个街区中脱颖而出。景观与时装的创意结合，打造了一个别样的绿色时尚世界。

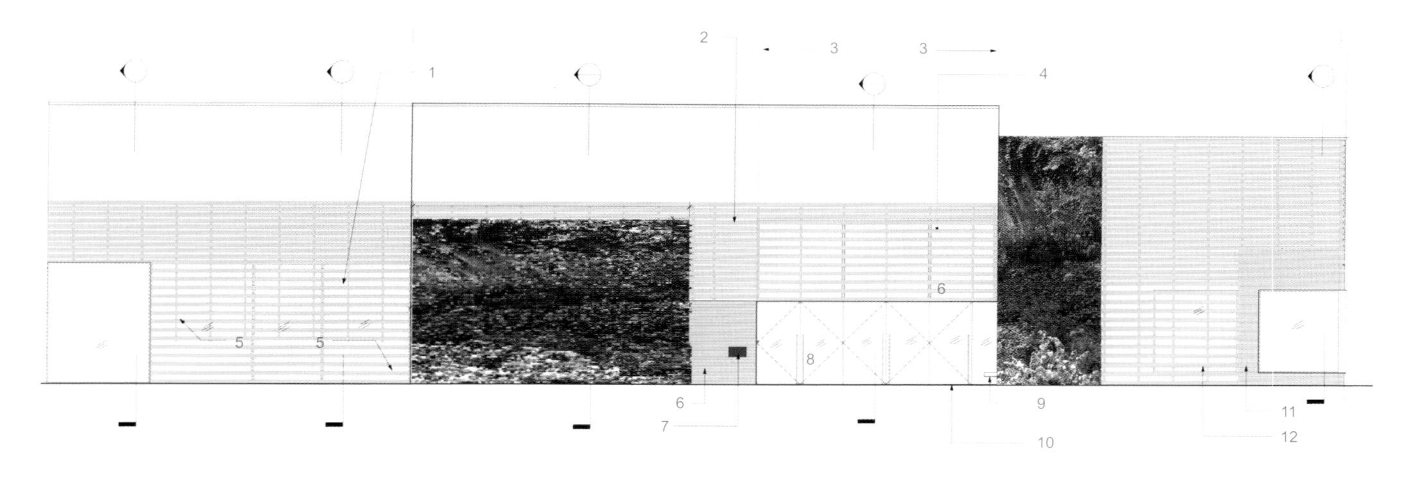

EAST ELEVATION

1. 6" wood boards hung on vertical 2x4 posts w/ 6" spacing o/glazing
2. 6" wood boards hung on vertical 2x4 posts w/ 2-1/2" spacing o/stucco facade
3. No wall sheathing on exterior wall in this area, exposed structure
4. Transom window beyond
5. Storefront window w/patterned laminated glazing behind louvres
6. Reclaimed barn wood horizontal siding
7. Laser-cut back-lit metal sign
8. Laminated glass
9. Reclaimed wood bench
10. Recessed entry beyond
11. Reclaimed barn wood horizontal panelling
12. Storefront window w/patterned laminated glazing behind louvres

东侧立面图

1. 木板（15.24厘米宽，纵向拼接，间隔15.24厘米，后面是玻璃幕墙）
2. 木板（15.24厘米宽，纵向拼接，间隔6.35厘米，后面是灰泥墙面）
3. 此处的外墙没有任何防护层，裸露的结构
4. 上方气窗
5. 店面橱窗（遮光栅格后面是夹层玻璃）
6. 回收利用的水平方向护墙板
7. 金属店招（激光切割，背光照明）
8. 夹层玻璃
9. 回收利用的木质长椅
10. 缩进式入口
11. 回收利用的水平木质贴面板
12. 店面橱窗（遮光栅格后面是夹层玻璃）

Brooks Avenue House

布克街别墅

Location:
Los Angeles, CA, USA
Architect:
Bricault Design
Photographer:
Kenji Arai & Danna Kinsky
Area:
352 sqm

项目地点：
美国，加利福尼亚州，洛杉矶
建筑设计：
布里科设计公司
摄影师：
新井健二、丹娜・金斯基
面积：
352平方米

Project description:

The clients for this project need more space to accommodate the needs of a growing family, but they were reluctant to leave their location in Venice – one of the few walkable neighbourhoods in Los Angeles. The solution was to maintain and remodel their existing 186-square-metre home, while creating a 158-square-metre addition and courtyard on the rear lane side.

With an ideal climate for much of the year, a primary design driver was to create a seamless connection between inside and outside, while eliminating the need for air conditioning. To this end, a central sculptural staircase links the ground floor with the rooftop deck, while doubling as a chimney to draw cooling breezes through the house. On the main floor (ground floor), a sequence of pivoting doors opens the house to the courtyard, while on the first floor, windows fold back and full-height exterior panels slide into walls. A system of cedar battens serve as a shading device along much of the addition.

The volume of the new master bedroom extends out from the first floor, creating a carport below. Its exterior is clad with a living wall system on three sides, visually tying together the courtyard greenery with the planted roof. All landscaping is fed with a combination of captured rainwater and recycled domestic grey water. The roof's softscape is divided between a highly productive vegetable garden and indigenous, low-maintenance grasses and shrubs. The roof also supports a solar panel array that is sufficient to meet household needs.

The house features a high-efficiency combination boiler, which supplies both radiant in-floor heating and domestic hot water. A hot water recirculation loop makes hot water available "on demand", while reducing consumption. Other features include low-flush toilets and non-toxic, low-VOC finishes, which are used throughout the house.

Technology:
Structure: Steel and wood
Heating: Hydronic radiant slab
Electricity: Grid-tied solar
Insulation: Icynene spray foam insulation, recycled cotton fibre battens

Green finishes & fixtures:
Cork and cork-rubber flooring, low-VOC paint, formaldehyde-free cabinetry, LED lighting (including retrofit of existing), dual-flush

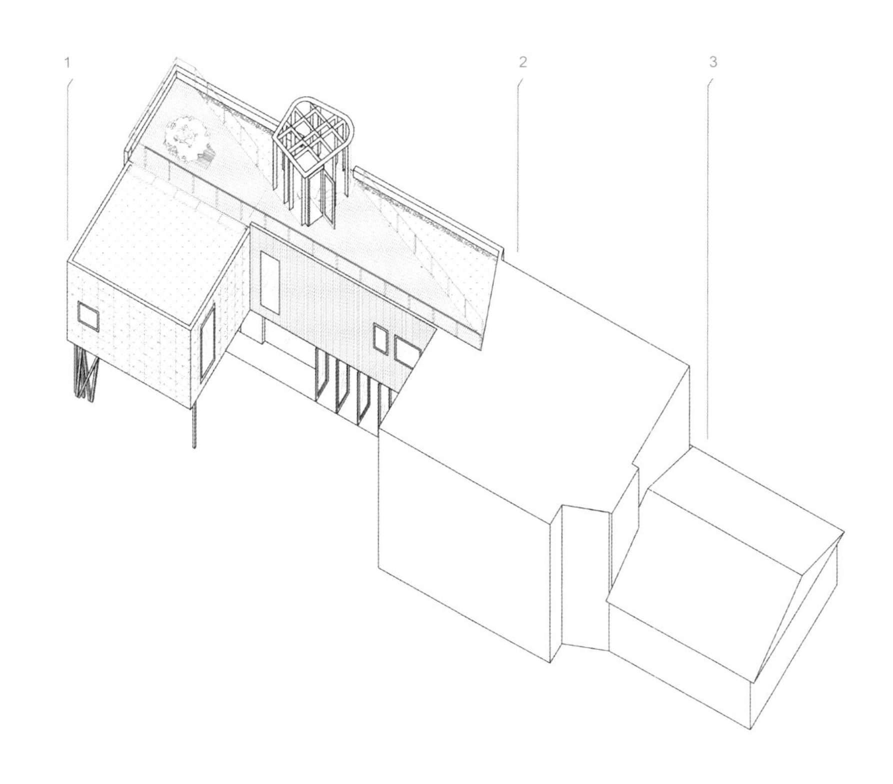

STRUCTURAL DRAWING
1. New addition
2. Second addition (1990s)
3. Original cottage (1940s)

结构示意图
1. 最新扩建部分
2. 20世纪90年代扩建的部分
3. 20世纪40年代的原建筑

本案是洛杉矶威尼斯区的一栋私人住宅。委托客户由于家庭人口增加，所以需要更大的空间，但又不愿搬家，因为威尼斯区是洛杉矶少数几个环境优美的街区之一。于是他们决定对老房子进行扩建。这栋别墅原来的建筑面积为186平方米，此次扩建了158平方米，还在屋后增加了庭院。

洛杉矶一年当中大部分时间都气候宜人，于是设计师决定突出室内外的通透性，免除空调的使用。为此，设计师在中央设置了一段造型别致的楼梯，从一楼直通屋顶平台。楼梯同时起到通风管道的作用，让凉风吹遍室内。一楼（即主楼层）安装了一排旋转门，让室内与庭院之间的联系更加紧密；二楼采用折叠窗，从楼面延伸到天花的板材嵌入墙内。扩建部分采用杉木栅栏作为遮阳屏障。

扩建部分的主卧位于二楼，伸出建筑主体外，下方空间用作车库。主卧的三面外墙进行了绿化，垂直绿墙成为绿色庭院与屋顶花园之间在视觉上的过渡。所有植物的灌溉都用收集的雨水和家庭污水解决。屋顶花园的景观可以分为两部分，一部分是菜园，种植的蔬菜可供家庭食用，另一部分是草皮和灌木，无需过多维护。屋顶上还有一系列太阳能板，收集的太阳能足够满足家庭所需。

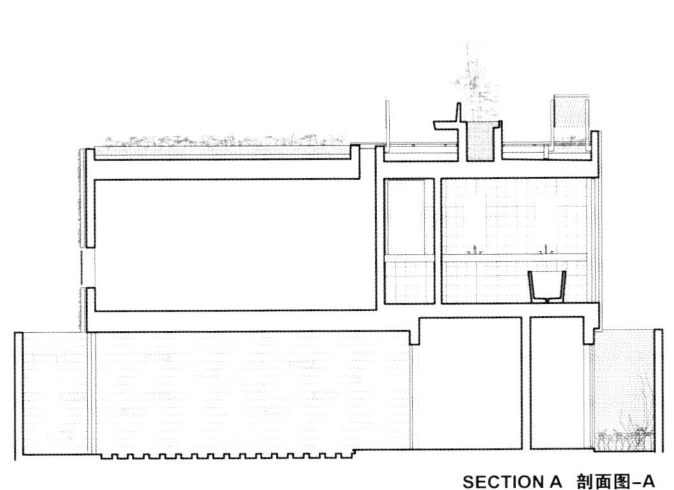

SECTION A 剖面图-A

SECTION B 剖面图-B

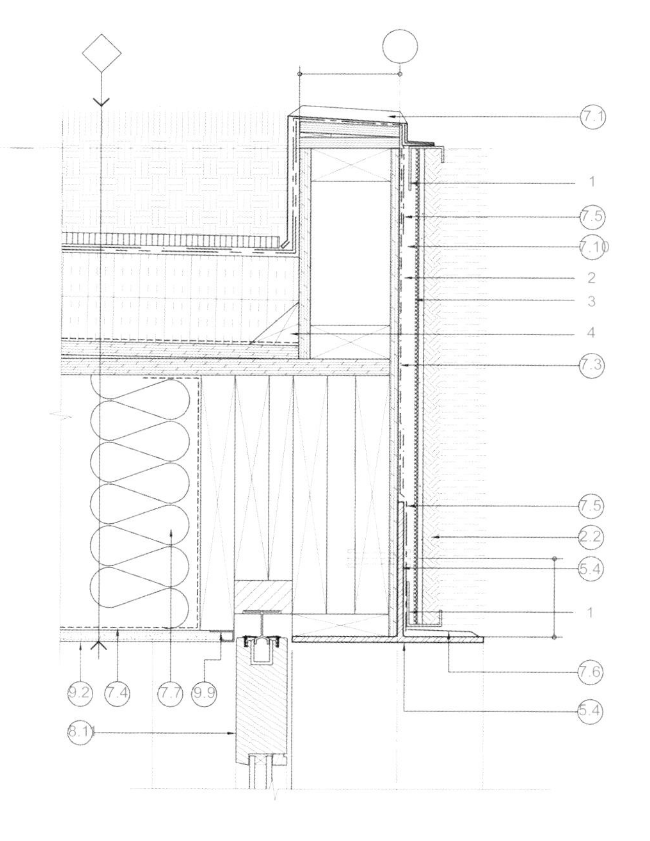

	Building paper	防潮纸
	Vapour barrier	蒸汽隔离层
	Self-adhered membrane	粘贴薄膜
	Dampproofing	防潮层
	Cap sheet	顶层板材
	Base sheet	底层板材
	Filter sheet	滤板
	Drainage mat	排水垫
	Plywood	胶合板

POCKET DOOR @ ROOF

1. Green wall frame galvanised steel
2. 5/8" plywood
3. Heavy felt on wire mesh
4. CANT strip

Material Legend:

2.2 - Growing medium
5.4 - Galvanised steel plate
7.1 - Zinc flashing
7.3 - 2 layers 30 minute building paper
7.5 - Self-adhered membrane
7.6 - Prefinished metal flashing
7.7 - Batt insulation
7.10 - 3/4" airspace
8.11 - Exterior pocket door
9.2 - Gypsum wall board, painted
9.9 - J-Mold

屋顶袖珍门

1.绿墙框架（镀锌钢）
2.胶合板（1.6厘米）
3.厚重型毛毡（铺设在金属丝网上）
4.垫瓦条

材料

2.2 – 植物生长介质
5.4 – 镀锌钢板
7.1 – 表面镀锌
7.3 – 两层防潮纸
7.5 – 粘贴薄膜
7.6 – 预制金属盖片
7.7 – 条毯式隔热层
7.10 – 气隙（1.9厘米）
8.11– 外部袖珍门
9.2 – 石膏板（上漆）
9.9 – “J”形模具

其他设计亮点还包括：高效能的复式锅炉，既能提供地暖所需的热能，又能供应家庭所需的热水；"热水循环网"的设计让热水可以随时供应，同时降低了能耗；另外还有卫生间冲水马桶的节水设计。室内装修全部采用无毒的、挥发性有机物含量低的材料。

技术信息：
结构：钢+木材
供暖：液体循环地热
供电：太阳能并网供电
隔热保温：喷涂泡沫绝缘层；回收利用的棉花纤维板条
绿色材料与家具用品：软木橡胶地面；挥发性有机物含量低的油漆；不含甲醛的家具；LED照明（包括对原有照明灯的改装）；双水路冲水马桶

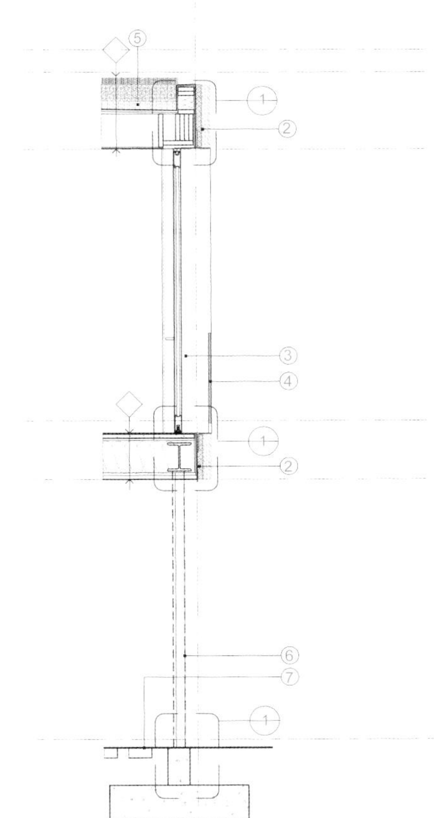

WALL SECTION
1. Roll-up garage door
2. Green living wall
3. Exterior pocket door
4. 36" high glass guardrail w/ceramic frit pattern
5. Rigid insulation
6. HSS - Structural
7. 3 1/2" noncombustible grasscrete asphaltic paving stones

墙壁剖面图
1. 车库卷帘门
2. 植物绿墙
3. 外部袖珍门
4. 玻璃扶手（彩釉玻璃，高92厘米）
5. 刚性绝缘
6. 高速钢结构
7. 不易燃的混凝土草格（9厘米）

Growing Up
生长的绿墙

Location:
Mexico City, Mexico
Architect:
VERDE VERTICAL ©
Client:
Park Plaza
Photographer:
Santiago Arau
Area:
750 sqm

项目地点：
墨西哥，墨西哥城
建筑设计：
“垂直绿化”设计公司
客户：
公园广场
摄影师：
圣地亚哥・阿劳
面积：
750平方米

Project Description:

This vertical garden with a lineal concept is the representation of an aerial view of one of the sections of the famous Xochimilco lake channels. This complements the integration of the garden with the concept of the hotel where it is located through the harmonic coexistence of its design and colours.

The incorporation of green tones and different textures turned the plaza where the garden is located into a nice space for the entrance, and arrival of pedestrians and vehicles that go to the hotel and adjacent commercial area.

This is one of the first vertical gardens made by VERDE VERTICAL ©, and it shows us the different tones of the year that make the garden change from the monotone into an icon of the west side of the city. This is how the architect Fernando Ortiz Monasterio Garza, director of VERDE VERTICAL ©, has achieved to create an excellent impact in a very short time by doing this work that accomplishes a symbiosis between urban structure and nature.

Interventions in the city are richer and more positive when they are in public spaces. Throughout different cities VERDE VERTICAL look for potential places that gather certain characteristics like: structure, water, and electrical supply. VERDE VERTICAL try to contact the owners of the different buildings to talk about the viability of the project. In the end this becomes a strategy to recover public space with interventions for the people who become the users and the ones who will get to see and enjoy them.

这面绿墙的设计采用线性结构，平面布局的灵感来自著名的索奇米尔科湖的一条河道的鸟瞰图。绿墙背后是一家酒店，设计师通过绿墙的设计手法和色彩，使其与酒店形成一个和谐的整体。

绿墙的存在让这个小广场焕然一新。茂盛的植被赋予空间别样的质感，深浅不一的绿色层次感分明。酒店入口因此成为一个令人向往的地方，不论是进出酒店的客人，还是去往附近商业区的过往行人和车辆，都会被绿墙所吸引。

本案是墨西哥“垂直绿化”设计公司最早打造的一批垂直花园之一。绿墙上的植被展示着一年四季的变化，把一成不变的单调环境变为生动的花园。这里已经成为墨西哥城西部的新地标。“垂直绿化”设计公司的建筑师费尔南多・奥尔蒂斯・莫纳斯特里奥・加尔萨负责本案的设计，他在很短的时间内实现了对城市环境成功的绿色改造，探索了城市与自然的共生关系。

相比私人空间，城市公共环境的设计更加丰富，起到更加积极的作用。“垂直绿化”设计公司在许多城市寻找适合开展垂直绿化设计的潜在空间，这样的空间需要具备如下元素：框架结构、水和供电。“垂直绿化”设计公司会与不同建筑的业主进行联系，探讨实施绿墙的可能性。最后，这样的过程就变成了“垂直绿化”公司用绿墙重塑公共空间的设计策略。他们的绿墙设计以明确的使用群体为目标，旨在用绿墙打造让使用者赏心悦目的环境。

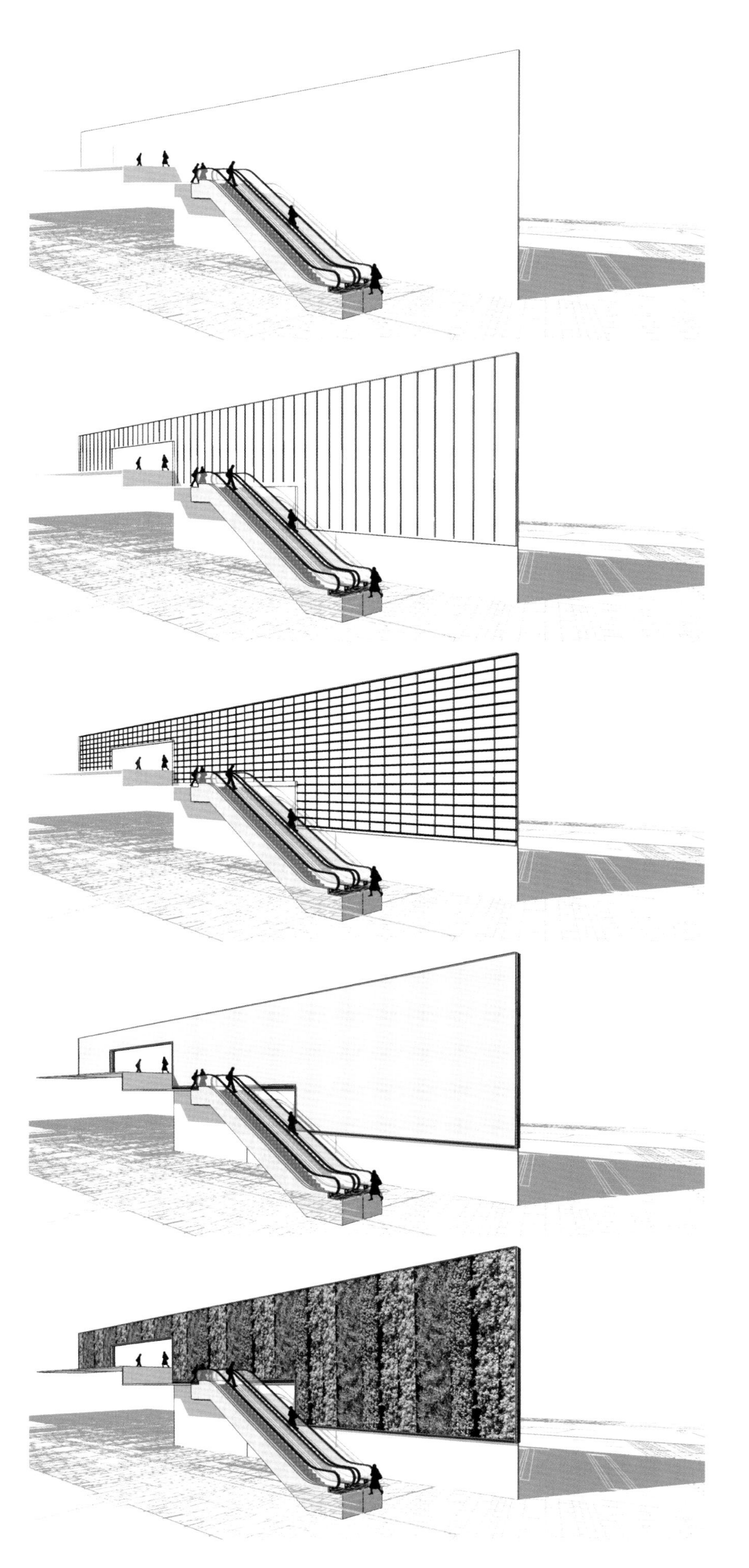

CONSTRUCTION PROCESS
施工过程示意图

Heineken House Mexico

墨西哥喜力啤酒屋

Completion date:
2011
Location:
Mexico City, Mexico
Architect:
Art Arquitectos
Lighting designer:
Noriegga Iluminadores Arquitectónicos Mexicanos
Photographer:
Paul Czitron & Marisol Paredes
Area:
1,300 sqm

竣工时间：
2011年
项目地点：
墨西哥，墨西哥城
建筑设计：
"艺术"建筑事务所
灯光设计：
墨西哥诺力加建筑照明事务所
摄影师：
保罗·齐特龙、马利索·帕雷德斯
面积：
1,300平方米

Project description:

The project for the Heineken bar and corporate offices of Cuauhtémoc Moctezuma Heineken Mexico located in a big house in the Polanco area of Mexico City – catalogued as historic patrimony by INBA – was selected through a competition organised by the client.

It was very important that the original and main architectonic features of the residence were preserved and at the same time incorporate, in a very contemporary trend, the brand image by creating the office areas and the Bar House for special guests and tastings. The main façade with classic ornaments remained untouched. The original partitions of the perimeter wall were partially covered with shade glass that gives privacy and shows a glimpse of the house architecture. A big copper marquee covered, with double glass in different shades of green and the logo, is on top a water mirror creating a reflecting and subtle brand presence at the entrance. The marquee also has on top a big green wall shaped as a bottle that is spilling foam all the way to the bottom. In order to have more room for the events, the parking area was covered with a curved shaped architectonic canvas creating an extra terrace. The paving was done with archeological stone that climbs up the walls and finishes in flower boxes.

On the top floor there is a private terrace with wood floors and glass screens covered with graphics that make the perfect ambiance for evening meeting and events. A roof garden was also done in this area for private council meetings, covering the unwanted views by the use of green curved walls and emphasising the spectacular ones with glass screens.

RADE
RK
Heineke
Moctezuma

ELEVATION 立面图

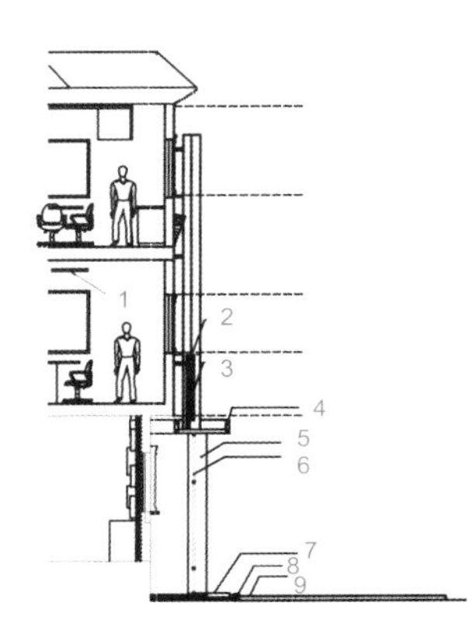

FAÇADE DETAIL

1. Alucobond overhead lights
2. 2" PTR metal structure lined with Durock
3. Plant panel
4. 3" metal framework lined with Durock
5. Metal column
6. Aluminium and glass
7. Water mirror
8. Edging
9. Grass

立面详图

1. 阿鲁克邦（Alucobond）顶灯
2. 金属结构，边缘采用杜力克材料（Durock）
3. 植物板
4. 金属框架结构，边缘采用杜力克材料
5. 金属柱
6. 铝材和玻璃
7. 水池
8. 边缘
9. 草地

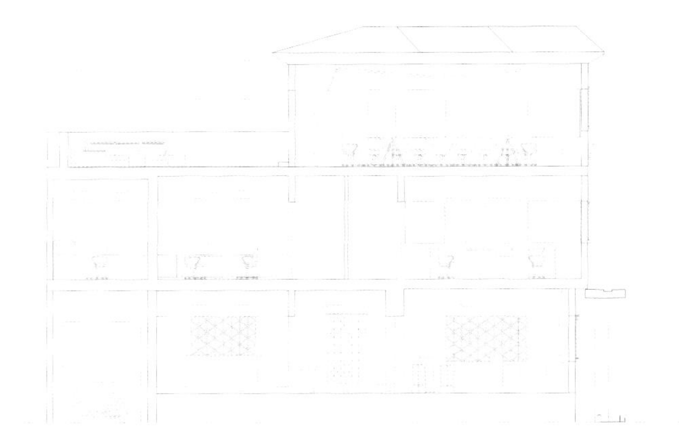

SECTION 剖面图

本案是喜力啤酒办公楼兼酒吧，坐落在墨西哥城波朗科区。这栋建筑物很古老，是当地的历史保护建筑。委托客户举办了竞赛，本设计案脱颖而出。

这座历史悠久的建筑物原是一栋住宅，设计师保留了原有的建筑特色，同时巧妙地加入喜力啤酒的品牌形象，呈现出古老与现代的有机结合。

建筑正门的外立面没有改动，保留了原来的古典装饰元素。四周的围墙增加了遮光玻璃，既保证了建筑内的私密性，又让人能从外面窥探建筑一斑。入口上方是大体量的铜板遮棚，搭配深浅不一的绿色双层玻璃和喜力品牌LOGO，门口还有倒影池，突出了品牌标识的存在感。铜板上方用植被塑造出啤酒瓶的造型，仿佛泡沫正从瓶口溢出，一直流到下面。为了有更多举办活动的空间，设计师在停车场上方加建了一个弧线造型的平台。平台上的铺装材料与墙壁和花池相同，都是古老的砖石。

三楼设置了一个私人平台，采用木质地板和玻璃屏风，玻璃上印有适合晚间会议或者举办活动的氛围的图案。平台上还有一个屋顶花园，用于理事会的私人会议。蜿蜒的绿色墙面遮挡住了不甚美观的视野，而视野好的方向则采用玻璃屏风。

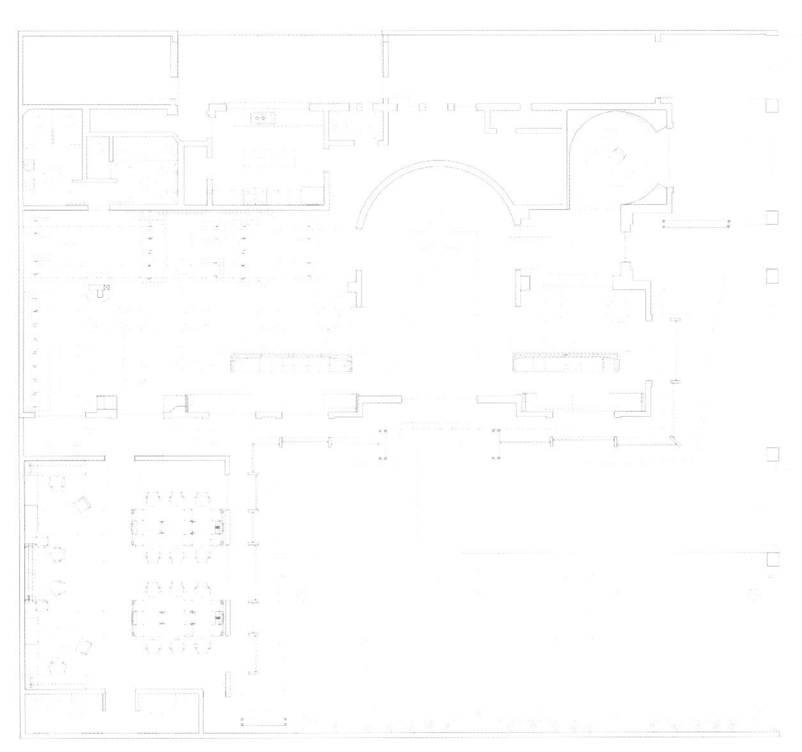

PLAN 平面图

Hotel Seven4One

741酒店

Completion date:
2013
Location:
Laguna Beach, California, USA
Architect:
Horst Architects
Photographer:
Aris Iliopulos
Area:
506 sqm

竣工时间：
2013年
项目地点：
美国，加利福尼亚州，拉古纳海滩
建筑设计：
霍斯特建筑事务所
摄影师：
艾瑞斯·伊利普洛斯
面积：
506平方米

Project description:

This thirteen-room boutique hotel in Laguna Beach, California, was recently renovated and remodelled by contemporary architect firm based in Laguna Beach, Horst Architects. The project reflects a forward-looking, contemporary design aesthetic while at the same time making a connection to the Laguna Beach cottages through its materials and two-storey exterior green wall. The material of recycled wood found on the exterior of the hotel has a weathered look to it, even showing worm holes, allowing the hotel to connect to the adjacent buildings, such as the historic Orange Inn which was established in 1931. The other material of Corten steel has an earthly quality to it, which channels the beautiful land and attitude of Laguna Beach residents. The exterior two-storey green wall also links to the green, open space that surrounds Laguna Beach, making it a distinctive community.

The fabulous new living walls on the exterior façade of the building and interior courtyard received a merit award from the Laguna Beach Beautification Council for "outstanding contribution to the beauty of Laguna Beach". Horst Architects' incredible design work helped generate what's essentially a gigantic piece of living artwork. It is believed that the walls provide a very distinguished, unique and eco-modern look that – by all accounts from passionate locals – is quintessentially "Laguna". Seven4one is meant to be a blank canvas enabling wedding couples to customise the space in countless different ways; this new living wall feature has stayed true to this objective while bringing a stunning and universally appreciable character of its own.

SEVEN4ONE
741

这家精品酒店位于加州拉古纳海滩，共有13间客房，2013年由本地的霍斯特建筑事务所进行了翻新。本案的设计体现出一种富有前瞻性的现代美学，同时，设计师利用材料和两层高的室外绿墙，在风格上贴近拉古纳海滩上的农舍式小别墅。酒店外墙的材料采用回收利用的木材，看上去历经风霜，甚至连上面的虫眼都清晰可见，这让这座酒店与周围的建筑能很好地融合，比如旁边的“橙色旅馆”，外观也很古旧，建于1931年。酒店外观的另一种主要材料是耐候钢，也极具质感，让这家酒店融入拉古纳海滩的美丽风景和当地居民的生活方式中。酒店外立面上两层高的绿墙把酒店与周围拉古纳海滩的绿色开放式空间联系起来，形成一个和谐而独特的整体环境。

酒店外立面上的绿墙和室内庭院的设计获得了拉古纳海滩环境美化委员会颁发的优秀奖，奖励其“为拉古纳海滩的环境所做的杰出贡献”。霍斯特建筑事务所用他们杰出的设计为拉古纳海滩打造了一件“活的艺术品”。绿墙的设计赋予这家酒店一种独特的、现代的、生态环保的外观。据当地人说，这一设计精准地体现了拉古纳海滩的精髓。741酒店秉承的理念是作一张白纸，让举办婚礼的新人自己用无数种不同的方法装饰他们的婚礼现场。这面绿墙也符合这一定位，同时自身也具备令人眼前一亮的、放之四海而皆准的独特美感。

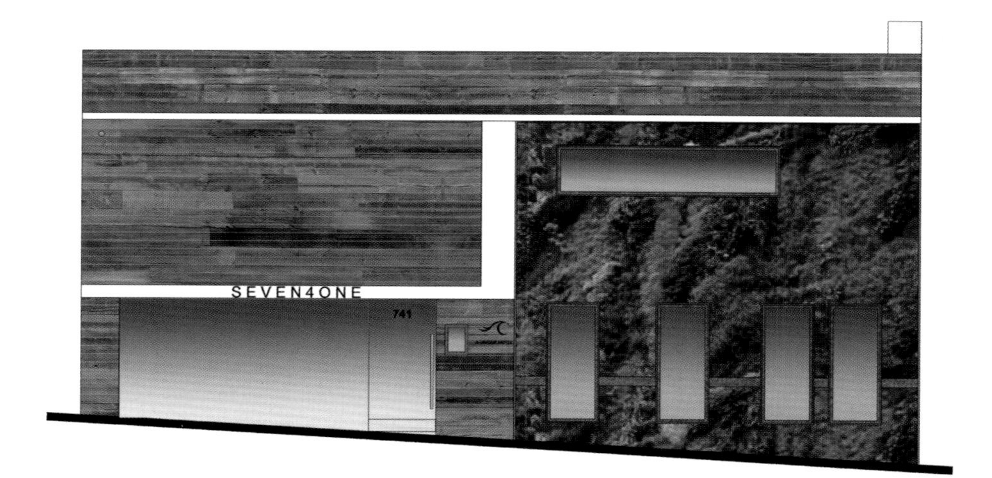

ELEVATION　立面图

PLAN

1. Room
2. Outdoor courtyard
3. Kitchen
4. Lounge
5. Lobby
6. Check-in

平面图

1. 客房
2. 户外庭院
3. 厨房
4. 休闲区
5. 大堂
6. 入口登记处

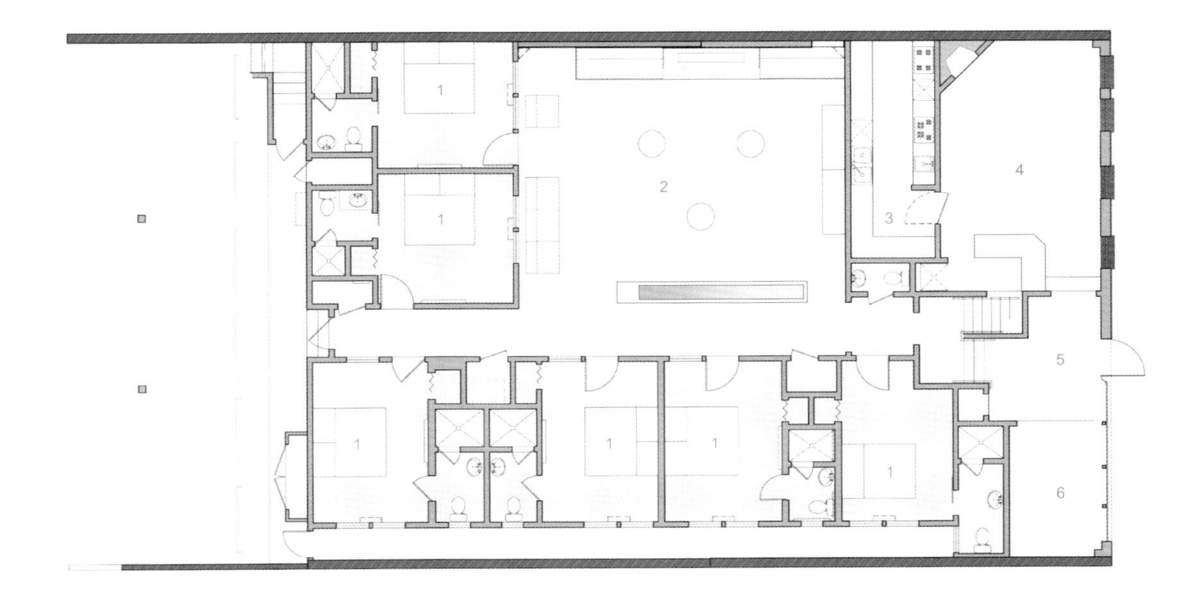

IDEO Morph 38

IDEO 38号公寓

Completion date:
2013
Location:
Bangkok, Thailand
Architect:
Somdoon Architects
Landscape architect:
Shma Company Limited
Interior designer:
Fix Design
Photographer:
W Workspace, Somdoon Architects, Spaceshift Studio
Area:
5,320 sqm

竣工时间：
2013年
项目地点：
泰国，曼谷
建筑设计：
Somdoon建筑事务所
景观设计：
Shma景观设计公司
室内设计：
菲克斯设计公司
摄影师：
W建筑摄影工作室、Somdoon建筑事务所、"空间转换"工作室
面积：
5,320平方米

Project description:

The scheme is located away from the high density and congestion of Sukhumvit road and into a blissfully green low-rise residential area. The development has been separated into two towers to maximise plot ratio, and each building targets to different potential tenants in character.

The two towers are visually interconnected through a folding "Tree Bark" envelope that wraps around from the 32-storey rear tower (Ashton) and 10-duplex-storey front tower (Skyle). This outer skin is a combination of precast concrete panels, expanded meshes and planters. The function of the skin varies from being sun shading devices to covering air condensing units. The bark on the west and east side strategically becomes green walls, in accordance to the tropic sun's orientation. The height of this wall is 65m on the front tower and 130m on the rear tower respectively, providing the residences and neighbours with a comfortable visual and natural environment.

Skyle is targeted for singles or young couples with the smallest unit footprint being 23.3 square metres. These duplex units are expressed vertically with variation of balconies and air condensing units.

In contrast, Ashton emphasises the horizontal

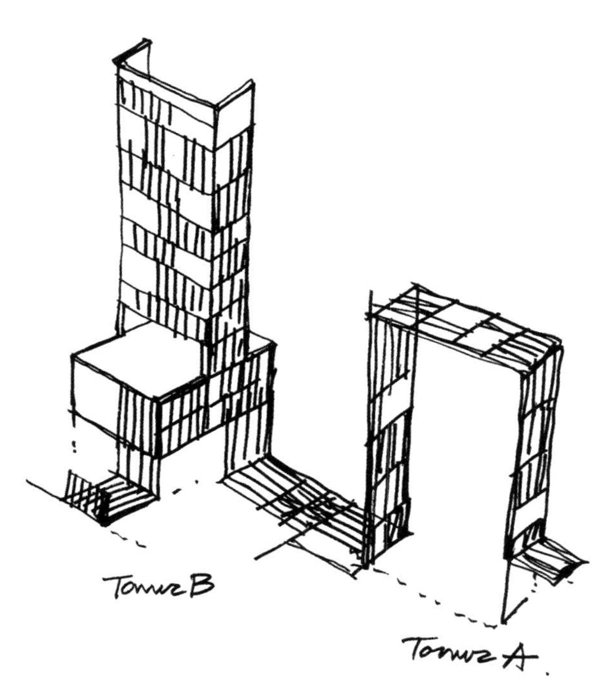

TREE BARK SKETCH “树皮”外墙手绘图

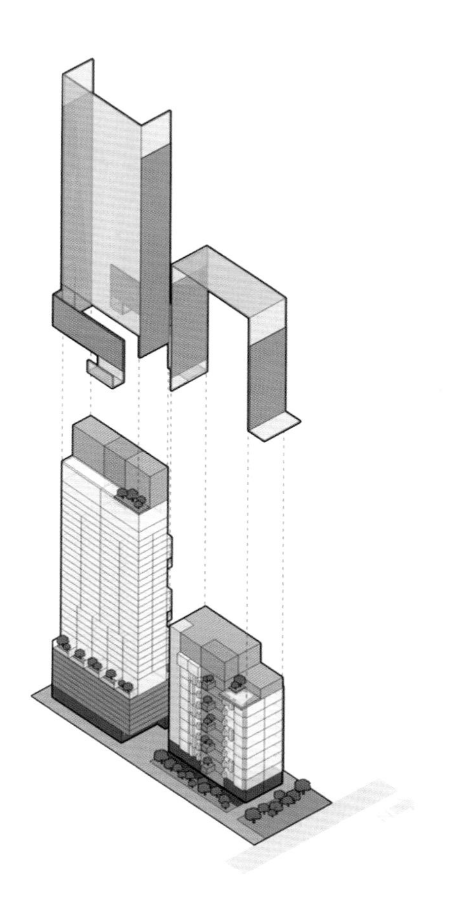

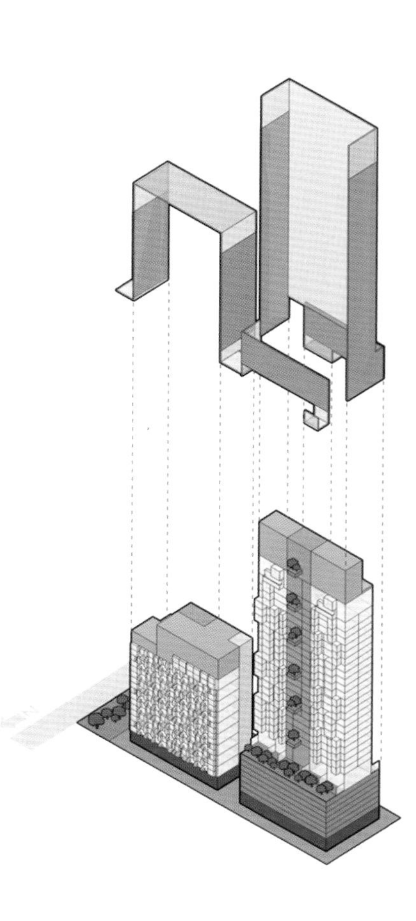

MASSING + TREE BARK
“树皮”外墙概念图

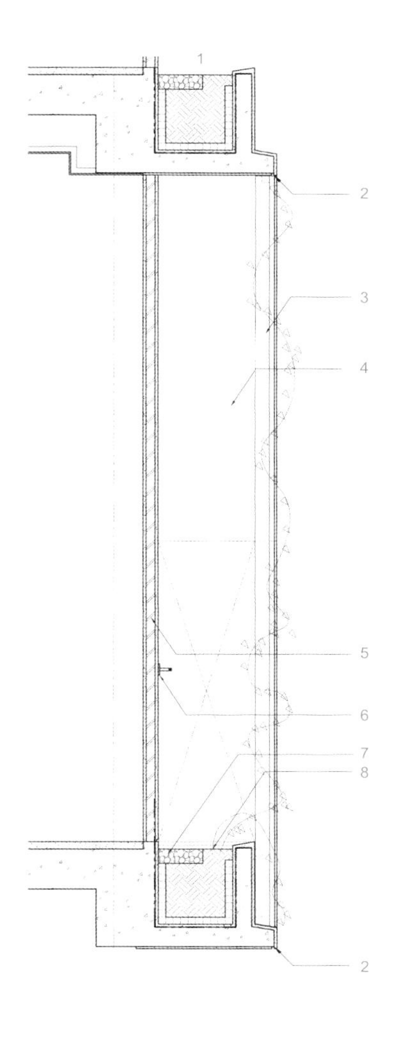

TREE BARK TYPICAL SECTION ON SKYLE
1. Planter
2. 15mm PVC grove
3. The tree bark: precast panel and aluminium expanded mesh panels
4. Cantilevered of unit mezzanine floor behind
5. External wall construction
6. Safety rail for planter service
7. Pebble for service way
8. Planter construction:
   a. Growing media
   b. Coarse sand
   c. Geotextile
   d. Subsoil drainage modules
   e. Waterproofing membrane
   f. Cement & screen to fall

"斯盖尔"楼树皮外墙标准剖面图
1. 种植槽
2. PVC材料（15毫米）
3. "树皮"：预制板和铝制网板
4. 后面夹层的悬臂
5. 外墙
6. 安全扶手（种植时用）
7. 种植通道上铺设鹅卵石
8. 种植槽结构：
   a. 植物生长介质
   b. 粗砂
   c. 土工织物
   d. 土壤下排水模块
   e. 防水薄膜
   f. 水泥

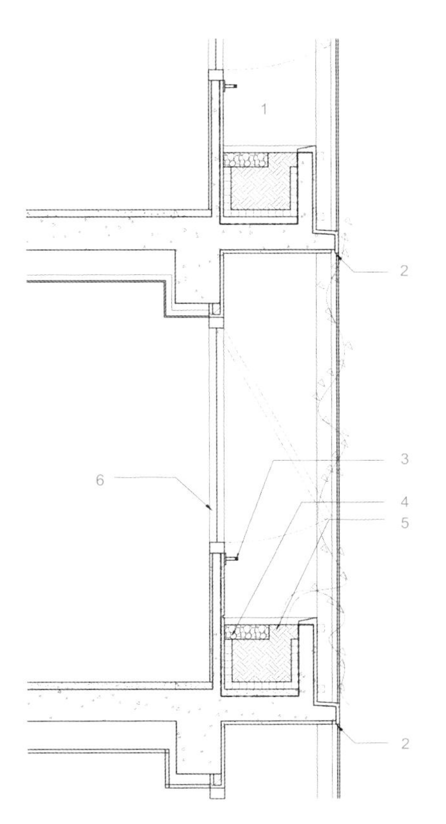

TREE BARK TYPICAL SECTION ON ASHTION
1. Planter
2. 15mm PVC grove
3. Safety rail for planter service
4. Pebble for service way
5. Planter construction:
   a. Growing media
   b. Coarse sand
   c. Geotextile
   d. Subsoil drainage modules
   e. Waterproofing membrane
   f. Cement & screen to fall
6. Top-hung aluminium window

"阿什顿"楼树皮外墙标准剖面图
1. 种植槽
2. PVC材料（15毫米）
3. 安全扶手（种植时用）
4. 种植通道上铺设鹅卵石
5. 种植槽结构：
   a. 植物生长介质
   b. 粗砂
   c. 土工织物
   d. 土壤下排水模块
   e. 防水薄膜
   f. 水泥
6. 顶悬式铝窗

and cantilevered spaces which are targeted to families. The unit sizes and types vary from a single bed with a reading room, to duplex units with a private swimming pool and a garden on 8th floor, and a four-bed duplex penthouse at top level. A 2.4m cantilevered living space projects from each unit on the north side. This is made up of a glazing enclosure on three sides providing the maximum view. Each unit on the south has a semi-outdoor balcony which is flexible in space. The double layer of sliding windows allow for a transition between a conventional balcony to an extended indoor living area.

The IDEO Morph 38 is a high-rise building which is located in stark contrast to the low-rise residential context. The project takes on a symbiotic relationship with the environment. Due to its sensitive design language and the green populated façade, the buildings have a natural aesthetic, making them a landmark and a pleasant environment for the neighbours and the city.

曼谷IDEO 38号公寓选址远离交通繁忙、人口密集的素坤逸路，转而寻求比较贴近自然的绿色环境，周围都是低矮的住宅楼。这个楼盘包括两栋高层建筑，以便实现土地最大化的利用，每栋建筑针对不同特点的潜在租户。

两栋建筑通过“树皮”状的外表实现了视觉上的统一。较高的一栋在后，32层，名为“阿什顿”楼；较低的一栋在前，10层（每层都是复式公寓），名为“斯盖尔”楼。“树皮”状外墙是在预制混凝土板材上设置种植网，栽种植物而成。这一层建筑表皮有多项功能，既能为建筑遮挡光线，又能遮盖住空气冷凝设备。西侧和东侧的“树皮”表层化身为绿墙，目的是阻挡热带炎热的阳光。前楼的绿墙高度为65米，后楼的绿墙高度为130米，为公寓住户和附近居民美化了环境，带来一股更加自然清新的气息。

“斯盖尔”大楼的目标住户是单身人士或者年轻夫妇，最小的公寓户型面积仅为23.3平方米。从外观上的阳台和空气冷凝设备就能看出这些复式公寓的区别，突出垂直的视觉效果。

“阿什顿”大楼则相反，侧重空间的水平分布，目标住户是大家庭。里面的公寓户型分为多种大小和类型，有带书房的单间，也有顶层的带阁楼的复式四居室。每户都从北面伸出2.4米长的悬臂空间，三面采用玻璃围墙，带来最佳视野。南侧的每户都有一个半户外式阳台，空间可以灵活使用，双层滑动窗的设计可以使其变身为室内起居室的延伸空间。

这两栋高层公寓建筑与周围低矮的住宅楼形成鲜明对比，设计上突出了建筑与环境二者之间的关系。微妙的设计语言，再加上绿色植物覆盖的外墙，让这两栋建筑具有一种别样的自然美感，打造了优美的自然环境，成为附近街区乃至整个曼谷市的新地标。

Illura Apartments
伊路拉公寓

Completion date:
2013
Location:
Melbourne, Australia
Architect:
Elenberg Fraser Architects
Landscape architect:
Tract
Design & Construct:
Fytogreen
Photographer:
Stuart Tyler
Area:
121 sqm

竣工时间：
2013年
项目地点：
澳大利亚，墨尔本
建筑设计：
埃伦贝格·弗雷泽建筑事务所
景观设计：
特拉克特建筑事务所
设计&施工：
菲多格林公司
摄影师：
斯图尔特·泰勒
面积：
121平方米

Project description:

Illura, the next stage in Melbourne's residential evolution.

Using nature as its key influence, Illura features organically shaped balconies and stunning landscaped elements, monumental vertical gardens with native grasses and plants that soar three storeys above you.

As a street view of a modern apartment block in an old part of inner Melbourne, the vertical garden makes a powerful impact. Made up of a series of four elevated sections with the gardens facing northeast, these will need to tolerate full sun in summer.

The architecture and the vertical gardens complement each other to create a modern and fresh look for the three-level apartments. Using strappy foliage and also flat-growing ground-covering plants the garden's textural elements will last as the garden evolves over time.

In nature, nothing is wasted. With this in mind, the designers have designed a living space that utilises both the finest natural materials and custom organic elements, allowing the spaces to flow effortlessly from interior to exterior.

The result is a continuous and efficient use of area, intelligently maximising the space and allowing for fluid transformation from one zone to another; hence the kitchen merges seamlessly with the dining and living area to create a natural rhythm of energy.

Illura's interiors are naturally warm and inviting. Generous room dimensions and a choice of 22 layouts allow you to create an environment that matches you individual style.

ELEVATION 立面图

NOTE

This is predominantly native vertical garden, reflecting Australian vegetation types, colours and textures.

The above species were selected for their ability to cope with the prevalent sun and wind exposures, and produce healthy growth & development on the vertical utilizing an appropriate, single, irrigation and fertigation approach.

Vertical swathes are employed to:
- Assist in limiting the development of cumulative shading
- To assist in controlling lateral wind movement
- Assist in producing limited humidity retentive microclimates between species
- Increase pest/disease control by limiting lateral pest spread, through separating species of different resistances

Each species is specifically positioned adjacent to compatible species & arrayed in such a manner so as to limit excessive shading & aggressive competition to the top, sides and base of each respective colony.

The growth form of each separate species has dictated its position in the wall, so as develop artificial ecologies, presenting long-term sustainability and low maintenance, form establishment through to the development of climax colony interactions.

SPECIES SELECTION

A = Agapanthus 'Peter Pan', lily, Lime green foliage, blue flowers x 285
B = Bulbine vagans, lily, light green foliage, yellow flowers x 103
C = Correa decombens, shrub, light green foliage, red/green flowers x 171
D = Dianella caerulea 'breeze', lily, green foliage, purple flowers x 438
E = Euonymus japonicas microphyllus, shrub, green foliage, insignificant flowers, compact shape x172
F = Lomandra confertifolia ssp. Rubiginosa 'frosty top', lily, blue/grey/green foliage, yellow scented flowers x 319
P = Pratia Pedunculata, groundcover, light green foliage, white flowers x 482
R = Correa reflexa, shrub, grey/green foliage, green, red or yellow-green flowers x173
T = Lomandra longifolia 'tankia', lily, lime green foliage, insignificant flowers x238
W = Lomandra confertifolia ssp. Rubiginosa 'wingarra', lily, green/ blue/grey foliage, yellow scented flowers x639
X = Casuarina glausa 'cosin it', pendulant shrub, bright green foliage producing interesting cascading effect x61
Z = Zieria prostrata 'carpet star', groundcover/small shrub, dark gren foliage, white flowers x 104
TOTAL 3185

备注

垂直花园内主要种植的是本地植物，展现了澳大利亚本土植被的类型、颜色和质地。所选植物都是能适应既定条件的物种（光照强烈、风力较大），在适当的灌溉和施肥条件下能够较好地生长。

绿墙表层植物的作用包括：
– 限制植被的遮阳效果
– 控制侧面风力
– 保持一定的湿度，创造适宜的微气候
– 将具有不同病虫害抵抗力的物种分开，通过限制边缘区害虫的传播，控制病虫害

每种植物种植的位置都是精心设计的，临近的植物都彼此相容，同时能够限制植物过度生长，以免过于遮挡阳光，或者在每个种植区顶部、底部和旁边形成过于激烈的生长竞争。

每种植物的生长形态决定了它在绿墙上的栽种位置，这样就形成了一种人工生态环境，确保长期可持续性以及易维护性。

植物品种

A = 非洲爱情花（百合属，叶石灰绿色，花蓝色，共285株）
B = 鳞芹（百合属，叶浅绿色，花黄色，共103株）
C = 科雷亚木（灌木，叶浅绿色，花红/绿色，共171株）
D = 桔梗兰（百合属，叶绿色，花紫色，共438株）
E = 小叶正木（灌木，叶绿色，花极细小，修剪整齐，共172株）
F = "霜顶"密叶新木姜子（百合属，叶蓝/灰/绿色，花黄色，味芳香，共319株）
P = 铜锤玉带草（铺地观赏植物，叶浅绿色，花白色，共482株）
R = 倒挂金钟（灌木，叶灰/绿色，花绿、红或黄绿色，共173株）
T = 长叶草（百合属，叶石灰绿色，花极细小，共238株）
W = "温加拉"密叶新木姜子（百合属，叶蓝/灰/绿色，花黄色，味芳香，共639株）
X = 木麻黄（灌木，叶翠绿色，如瀑布般倾泻而下的效果，共61株）
Z = 澳大利亚芸香（又名"星星毯"，铺地观赏植物/小灌木，叶深绿色，花白色，共104株）
植物共计3185株

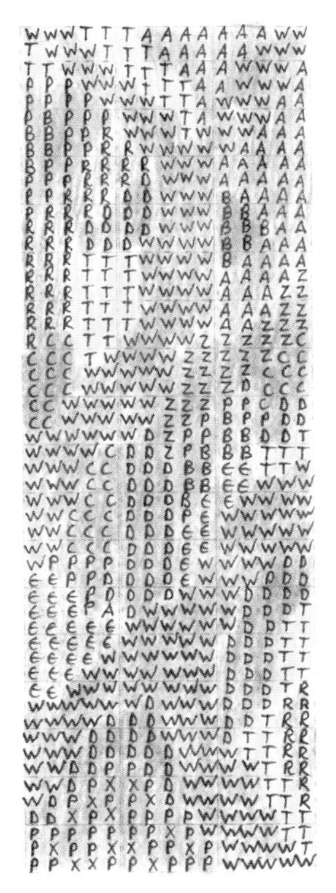

1

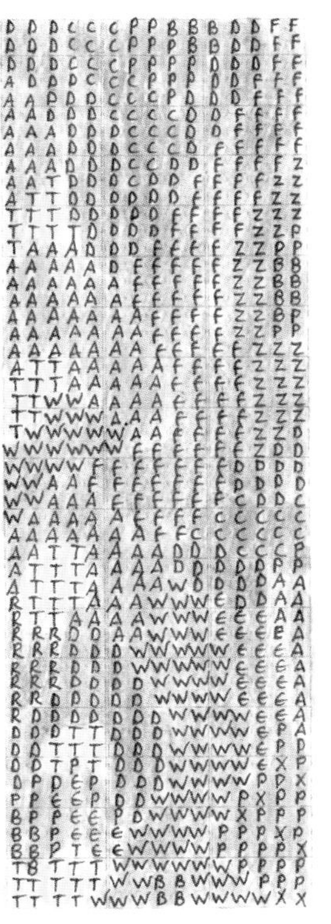

2

3

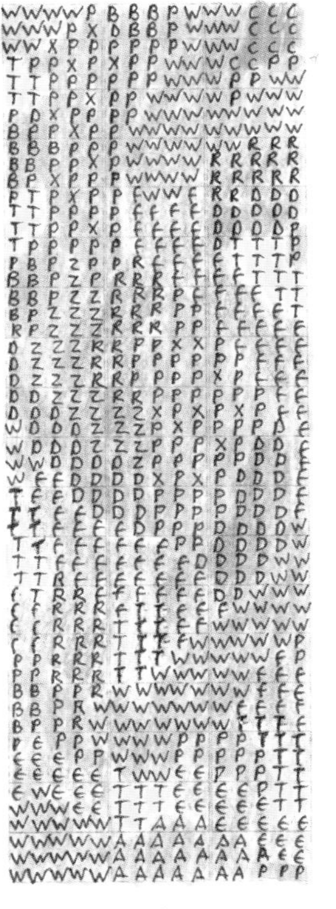

4

PLANTING PLAN

1. Garden 1
2. Garden 2
3. Garden 3
4. Garden 4

垂直花园种植平面图

1. 花园1
2. 花园2
3. 花园3
4. 花园4

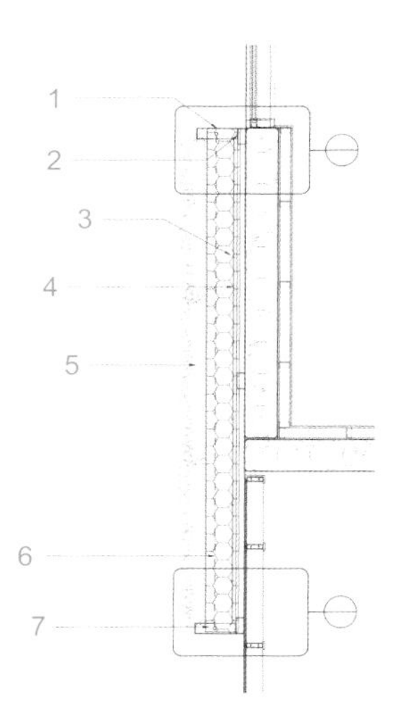

TYPICAL VERTICAL LANDSCAPE WALL PLAN

1. Powder-coated aluminium sheet cladding on light steel framing as required to galvanised sub frame
2. 100×50 RHS steel member fixed to the wall
3. Stainless steel fixing hooks TEK screwed to FC sheet substrate
4. 22mm painted (black) FC sheet substrate fixed over steel frame
5. SS drainage tray at base of panel
6. Irrigation drip line
7. Irrigation feeder pipe

垂直绿墙标准平面图

1. 镀膜铝板（镶嵌在轻型镀锌钢质框架中）
2. 钢构件（100毫米×50毫米矩形钢管，固定在墙上）
3. 不锈钢固定挂钩（用螺丝固定在纤维水泥板上）
4. 纤维水泥板（22毫米，表面上黑色漆，固定在钢质框架上方）
5. 嵌板底部的不锈钢排水托盘
6. 灌溉滴水线
7. 灌溉供给管

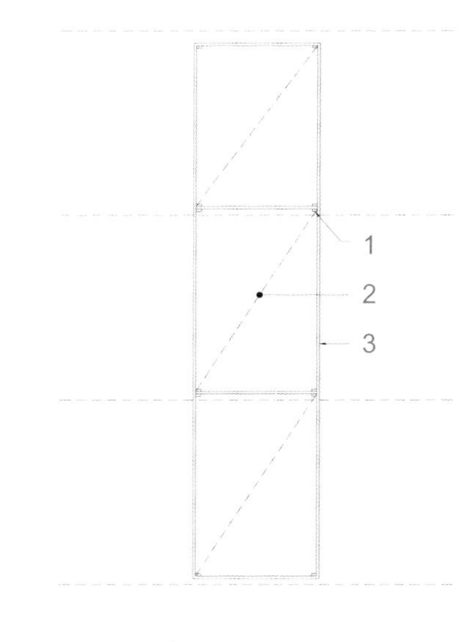

WIRE MESH LANDSCAPE SCREEN ELEVATION

1. 120×120×6 black powder-coated wall-mounted base plate
2. Black powder-coated wire mesh fixed to wall-mounted frame
3. 40×40 SHS black powder-coated wall-mounted framing members

金属丝网景观墙立面图

1. 黑色镀膜底座（120毫米×120毫米×6毫米，安装在墙上）
2. 黑色镀膜金属丝网（固定在框架结构上，框架结构安装在墙上）
3. 黑色镀膜薄钢框架（40毫米×40毫米，安装在墙上）

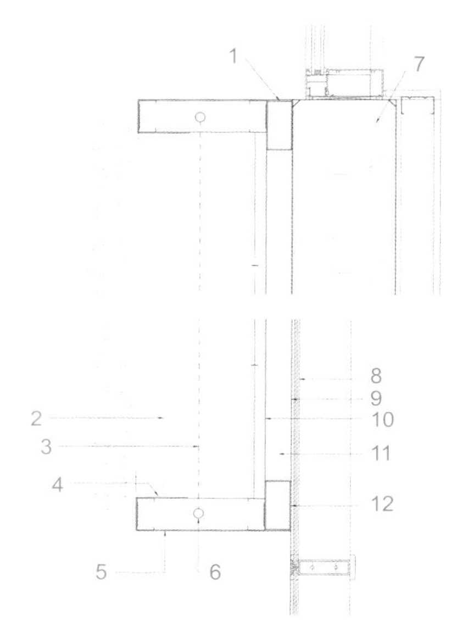

TYPICAL VERTICAL LANDSCAPE WALL PLAN DETAIL

1. 100×50 RHS black powder-coated steel member fixed to precast concrete wall panel
2. Fytowall module
3. Irrigation drip line
4. Provide 100mm deep return of poweder-coated aluminium sheet cladding to inside face of perimeter frame
5. Powder-coated aluminium sheet cladding on light steel framing as required to galvanised sub frame
6. Irrigation feeder pipe
7. Precast concrete wall
8. BAL1 glass balustrade
9. Concrete slab edge
10. 22mm painted (black) FC sheet substrate fixed over steel frame
11. 50×50 SHS black powder-coated steel members welded between 100×50 RHS at selected centers
12. 100×50 RHS black powder-coated steel member fixed to slab edge

垂直绿墙标准平面大样图

1. 黑色镀膜钢构件（100毫米×50毫米，固定在预制混凝土墙板上）
2. Fytowall模件
3. 灌溉滴水线
4. 镀膜铝板（贴在边缘框架的内表面）
5. 镀膜铝板（镶嵌在轻型镀锌钢质框架中）
6. 灌溉供给管
7. 预制混凝土墙
8. 玻璃扶栏
9. 混凝土板边缘
10. 纤维水泥板（22毫米，表面上黑色漆，固定在钢质框架上方）
11. 黑色镀膜薄钢构件（50毫米×50毫米，焊接在100毫米×50毫米矩形钢管之间选定的中心点上）
12. 黑色镀膜矩形钢管（100毫米×50毫米，固定在混凝土板边缘）

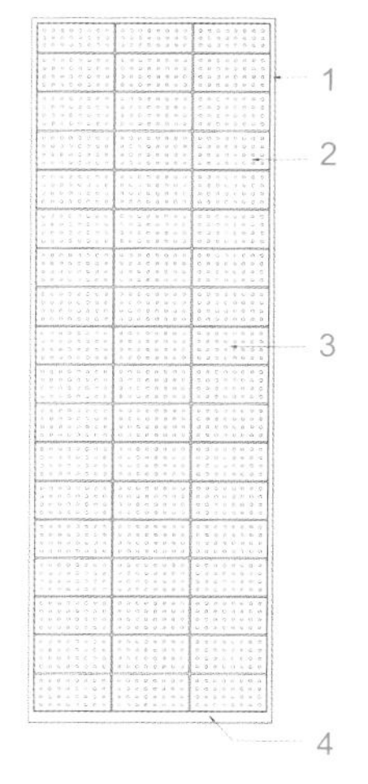

VERTICAL LANDSCAPE WALL FRAMING ELEVATION

1. 100×50 RHS black powder-coated steel members
2. 100×50 RHS black powder-coated steel members to be fixed to slab edges
3. 22mm FC sheet (black paint) fixed over steel frame
4. 50×50 SHS black powder-coated steel members fixed to 100×50 RHS's

垂直绿墙框架立面图

1. 黑色镀膜矩形钢管（100毫米×50毫米）
2. 黑色镀膜矩形钢管（100毫米×50毫米，固定在混凝土板边缘）
3. 纤维水泥板（22毫米，上黑色漆，固定在钢质框架上方）
4. 黑色镀膜薄钢构件（50毫米×50毫米，固定在100毫米×50毫米矩形钢管上）

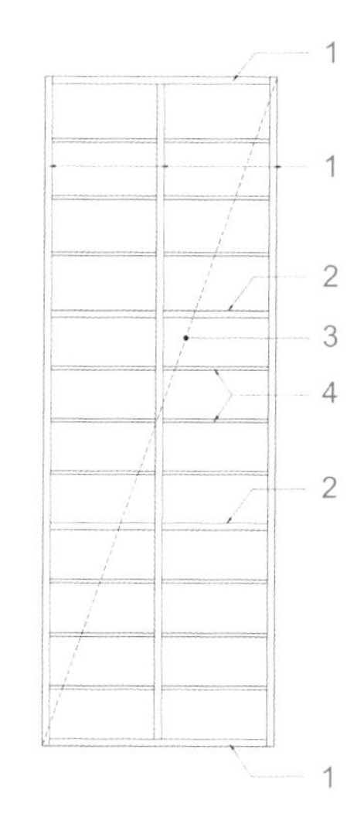

VERTICAL LANDSCAPE WALL ELEVATION

1. Poweder-coated aluminium sheet cladding on light steel framing as required to galvanised sub frame
2. Stainless steel fixing hooks (dashed) TEK screwed to FC sheet substrate
3. 1000×500 Fytowall modules
4. Drainage gutter in base of frame (dashed)

垂直绿墙立面图

1. 镀膜铝板（镶嵌在轻型镀锌钢质框架上）
2. 不锈钢固定挂钩（虚线，用螺丝固定在纤维水泥板上）
3. Fytowall模件（1000毫米×500毫米）
4. 框架底部的排水槽（虚线）

A natural palette of timber and stone encourages you to live within the landscape. Full timber joinery is designed to wrap the walls to form ingenious study nooks. Solid charcoal stone smoothly transforms from island counter to dining bench.

Double-glazed full-length windows offer the ultimate in sustainability and privacy. When the mood suits, they can open to create a terrace that emerges naturally from your living space.

Everything has been thoughtfully crafted to produce the perfect city retreat.

伊路拉公寓代表了墨尔本的住宅发展进入了一个新阶段。

伊路拉公寓的外观侧重自然元素。除了有机形态的阳台之外，公寓大楼的外墙进行了垂直绿化，种植的都是本地植物，植被在三层高的外立面上蔓延生长，非常引人注目。

这座现代公寓位于墨尔本中心的老城区，大楼表面的“垂直花园”成为街景的一部分，极具视觉冲击力。“垂直花园”由四个部分组成，朝向东北方向。夏季，这些植物需要经受墨尔本骄阳的考验。

在这栋三层高的公寓大楼中，建筑与绿化相辅相成，共同呈现出现代的、充满朝气的外观形象。绿墙上不仅有在表面生长的表层植物，也有具备观赏价值的大叶植物，呈现出层次分明的质感。随着植物逐渐生长，“垂直花园”将展现不同的面貌。

大自然中没有浪费。设计师希望能把这一点借鉴到他们的设计中。他们采用质量精良的天然材料和有机形态的造型，让室内空间自然地延展到室外。

通过这种设计手法，设计师实现了对空间连续性的、高效的利用，巧妙地在各个空间之间实现了自然的过渡。比如厨房与餐厅和起居室相连，没有明显的空间界限，创造出自然的空间韵律。

伊路拉公寓的室内空间也呈现出自然、温暖的氛围。房间很宽敞，有22种户型可供选择，一定可以营造出符合你个人风格的空间环境。

天然的建筑材料（如木材和石材）与外墙的绿化相互呼应。凹陷处的墙面采用天然木板贴饰，营造出适合学习的安静角落。桌台和长椅采用坚固的石材。

双层玻璃的落地窗既能拉近室内外的距离，同时又确保了室内的私密性。如果需要的话，可以把窗子完全打开，形成一个露台，与起居室连成一体。

总之，这栋公寓的每一个细节都是设计师精心打造的，目标就是在城市环境中创造一个完美的世外桃源。

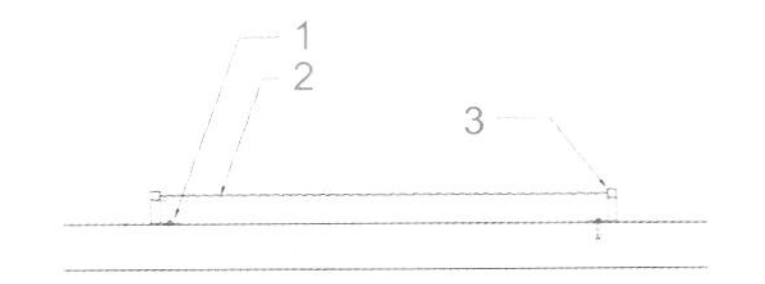

WIRE MESH LANDSCAPE SCREEN ELEVATION

1. 120×120×6 black powder-coated wall-mounted base plate
2. Black powder-coated wire mesh fixed to wall-mounted frame
3. 40×40 SHS black powder-coated wall-mounted framing members

金属丝网景观墙平面图

1. 黑色镀膜底座（120毫米×120毫米×6毫米，安装在墙上）
2. 黑色镀膜金属丝网（固定在框架结构上，框架结构安装在墙上）
3. 黑色镀膜薄钢框架（40毫米×40毫米，安装在墙上）

TYPICAL VERTICAL LANDSCAPE WALL ELEVATION (NORTH)

1. Vertical landscaped wall

垂直绿墙标准立面图（北侧）

1. 垂直绿墙

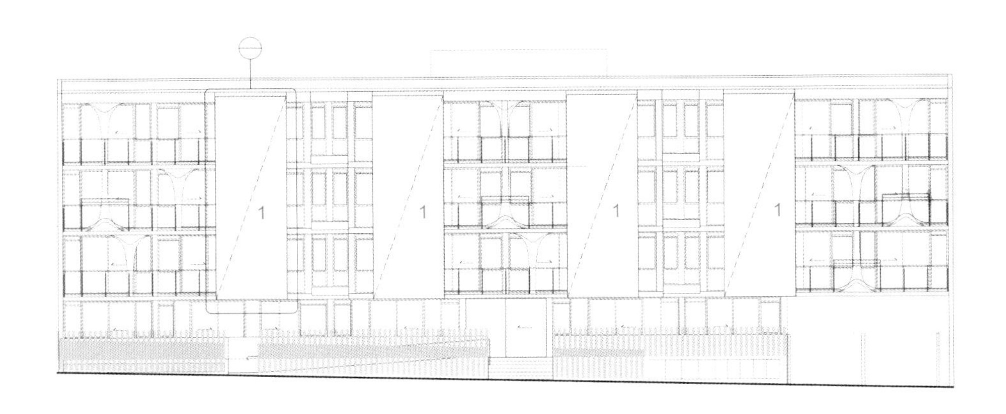

Living wall at the Rubens Hotel at the Palace

伦敦皇宫鲁宾斯酒店绿墙

Completion date:
2013
Location:
London, UK
Principal architect:
Gary Grant of the Green Roof Consultancy
Supplier and installer:
Treebox Ltd
Photographer:
Jane Dickson
Area:
350 sqm (living wall) + 125 sqm (living louvres)

竣工时间：
2013年
项目地点：
英国，伦敦
建筑设计：
绿色屋顶咨询公司，加里·格兰特
绿墙供应商及施工方：
Treebox 有限公司
摄影师：
简·迪克逊
面积：
350平方米（绿墙）+125平方米（百叶板绿化）

Project description:

Standing at 350 square metres and with a total of 10,000 herbaceous plants and ferns, the Living Wall at the Rubens at the Palace Hotel will be one of the largest and most environmentally beneficial living walls in London, boasting a number of unique innovations.

On the doorstep of Buckingham Palace, and designed to provide waves of blossoming plants throughout the year which will create bands of colour across the wall, it will also be one of London's most visually prominent and colourful living walls, brightening the popular tourist walk from Victoria station to the Royal residence and improving the air quality for those living and working in the area.

The living wall will provide wildlife habitat, help keep the hotel cooler in summer and warmer in winter, clean the air, deaden noise and bring more cheer to this corner of Victoria, which is presently undergoing significant renovation.

The plant list includes a wide variety of native species and those deemed by the Royal Horticultural Society as attracting insect pollinators – drawing bees, butterflies and birds, which is crucial in light of the decline in the bee population and the government's recent announcement for an urgent review. Buttercups, two varieties of crocus and strawberries will particularly attract butterflies and bees. From spring bulbs to evergreen geraniums, the plants have been selected with seasonal colour in mind, ranging from blues, pinks, purples, whites and yellows.

The wall will also improve the air quality in

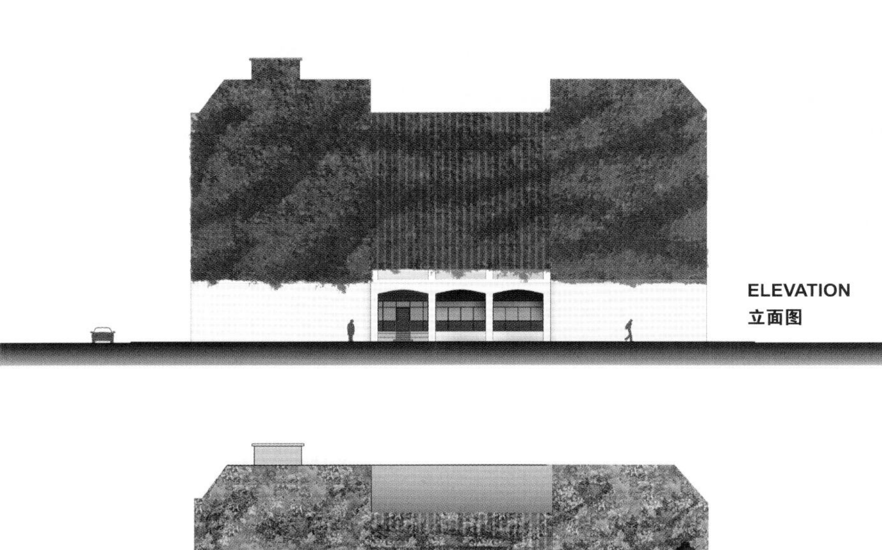

英国伦敦的皇宫酒店绿墙面积覆盖350平方米，种植有10,000多株草本植物和蕨类植物，是伦敦面积最大、最有利于环境健康的绿墙之一，在设计上有多项创新。

皇宫酒店位于白金汉宫对面，绿墙的设计旨在用各种颜色的植被来装点外立面，让酒店外观一年四季都呈现出生机与活力。这是伦敦最具视觉冲击力、最色彩斑斓的绿墙之一，它的存在让从维多利亚车站到皇家宅邸这一段著名的旅游景点更加丰富多彩，同时它还能改善空气质量，为在附近工作、生活的人们创造更加健康的生活环境。

皇宫酒店的绿墙有多项功能：提供野生生物栖息地；让酒店内保持冬暖夏凉；净化空气；隔音降噪；为该街区注入活力（这一街区正在进行大规模的重建）。

绿墙选用的植物包括许多品种的本地植物，还有皇家园艺协会指定的能够吸引传粉昆虫（包括蜜蜂、蝴蝶和鸟类）的植物；这一点尤为重要，因为这一地区的蜜蜂数量正在下降，当地政府刚刚宣布要采取紧急行动。金凤花、番红花和草莓尤其能够吸引蝴蝶和蜜蜂。从春季开花的鳞茎类植物，到四季常绿的天竺葵，设计师在植物的选择上特别注意色彩的变化，随着季节变换，外墙会呈现出蓝色、粉色、紫色、白色和黄色。

the area. The vegetation can trap microscopic pollutants known as particulate matter (PM10s). High levels of PM10s have been shown to cause respiratory illnesses that can result in death, so keeping their levels down is important for the health of the surrounding population.

The wall also includes a number of special features, including irrigation tanks to store rainwater harvested from the roofs. In a world's first, the wall will also be programmed to respond to heavy rain by slowly pumping water through the wall and increasing storage capacity of the rainwater tanks, taking the pressure of the drains and thereby helping to reduce the risk of surface water flooding in the neighbourhood.

Armando Raish, managing director of Treebox, said: "Due to the variety of plants used in its construction, we expect the living wall at The Rubens to significantly increase the number and variety of bugs and bees in this part of Victoria, helping to promote biodiversity and return nature to this urban environment. The wall will also help improve the respiratory health of the people who live and visit Victoria by absorbing pollutants, an important feature of the wall given the mounting evidence that shows just how harmful particulate matter can be to human health."

Features:
• Total of 10,800 herbaceous plants and ferns
• Total of 1,900 bulbs and flowering plants
• 23 different species of plants including natives and plants that will attract pollinators
• The water absorption capacity for a single panel is 35-40% of the volume of the soil equivalent to 7,500 litres of water for the Rubens hotel installation
• Uses two storage tanks to collect rainwater for irrigation
• 10,000 litres of rainwater harvesting capacity in the storage tanks
• Wall supports 16 tonnes of soil
• 6 tonnes of living wall system
• Easiwall pro living wall system manufactured in the UK out of recycled polypropylene developed by Treebox Ltd
• Plants have been selected with seasonal interest in mind, ranging from blues, pinks, purple, whites and yellows as well as the herbaceous plants

Aquilegia vulgaris
耧斗菜

Asplenium scolopendrium
美洲蕨

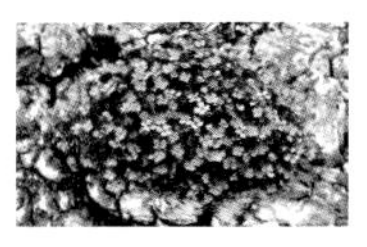
Aubretia deltoidea
匙叶南庭荠

Crocus sp
番红花

Crocus sp
番红花

Deschampsia flexuosa
曲芒发草

Fragaria vesca
野草莓

Galanthus nivalis
雪花莲

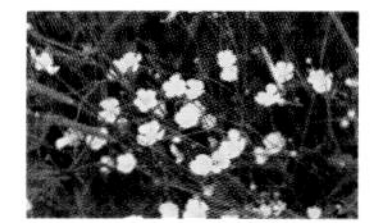
Ranunculus acris
金凤花

Silene dioicia
红色剪秋萝

Scientific name 拉丁语学名	**Common name** 俗称	**Native** 本地植物	**RHS pollinator** 传粉植物	**Color of flower** 花卉颜色	**Evergreen** 常绿植物
Aquilegia vulgaris	Columbine 耧斗菜	Yes 是	Yes 是	White/Blue 白/蓝	No 否
Asplenium scolopendrium	Hart's tongue fern 美洲蕨	Yes 是	No 否	/	Yes 是
Aubretia deltoidea	Aubretia 匙叶南庭荠	No 否	Yes 是	Blue 蓝	No 否
Crocus sp	Crocus 番红花	No 否	Yes 是	Yellow 黄	No 否
Crocus sp	Crocus 番红花	No 否	Yes 是	Blue/White 蓝/白	No 否
Deschampsia flexuosa	Tufted hair grass 曲芒发草	Yes 是	/	/	Yes 是
Fragaria vesca	Strawberry 野草莓	No 否	Yes 是	White 白	No 否
Galanthus nivalis	Snowdrop 雪花莲	Yes 是	Yes 是	White 白	No 否
Geranium macrorrhizum	Balkan Cranesbill 巨根老鹳草	No 否	Yes 是	Pink 粉	/
Hedera helix	Ivy 常春藤	Yes 是	Yes 是	Green 绿	Yes 是
Hebe pinguifolia	Hebe 长阶花	No 否	Yes 是	Blue 蓝	Yes 是
Onobrychis viciifolia	Sainfoin 红豆草	Yes 是	No 否	Pink 粉	No 否
Polystichum tsus-simense	Korean rock fern 对马耳蕨	No 否	Yes 是	/	Yes 是
Potentilla fructosa	Shrubby cinquefoil 金露梅	No 否	Yes 是	Yellow 黄	No 否
Primula vulgaris	Primrose 欧洲报春花	Yes 是	/	Yellow 黄	No 否
Ranunculus acris	Buttercup 金凤花	Yes 是	No 否	Yellow 黄	Yes 是
Silene dioicia	Red campion 红色剪秋萝	Yes 是	No 否	Pink 粉	No 否
Stellaria holostea	Stichwort 繁缕花	Yes 是	No 否	White 白	No 否
Vinca minor	Periwinkle 小蔓长春花	No 否	Yes 是	Blue 蓝	Yes 是
Violia riviniana	Dog violet 紫罗兰	Yes 是	No 否	Blue 蓝	No 否

Geranium macrorrhizum
巨根老鹳草

Hebe pinguifolia
长阶花

Hedera helix
常春藤

Onobrychis viciifolia
红豆草

Polystichum tsus-simense
对马耳蕨

Potentilla fructosa
金露梅

Primula vulgaris
欧洲报春花

Violia riviniana
紫罗兰

Vinca minor
小蔓长春花

Stellaria holostea
繁缕花

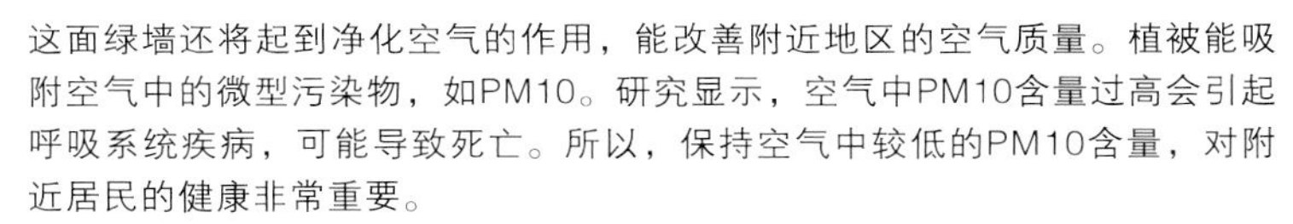

这面绿墙还将起到净化空气的作用，能改善附近地区的空气质量。植被能吸附空气中的微型污染物，如PM10。研究显示，空气中PM10含量过高会引起呼吸系统疾病，可能导致死亡。所以，保持空气中较低的PM10含量，对附近居民的健康非常重要。

皇宫酒店的绿墙还有许多特色，比如灌溉设计（从屋顶收集雨水，储存起来用作灌溉）。设计师采用了世界首创的技术，利用绿墙解决强降雨带来的问题。他们加大了雨水收集箱的储水容量，并让雨水缓慢地对绿墙进行灌溉。这样一来，缓解了排水管道的压力，因此也降低了附近街区街道雨水泛滥的危险。

英国绿墙设备供应商Treebox的总经理阿曼多・莱施表示："由于施工中用到多种植物，所以我们有理由预期，这面绿墙将会增加附近街区的虫子和蜜蜂的数量和种类，有助于建立生物多样性，让大自然回归到城市环境中来。这面绿墙还能改善人们的呼吸系统健康，因为它能够吸附污染物。越来越多的证据显示，空气中的污染物对人类健康会产生多么大的危害，所以绿墙的这项功能尤为重要。"

设计特色：

- 10,800株草本植物和蕨类植物
- 1,900株鳞茎类植物和开花植物
- 23个植物品种，包括本地植物和能够吸引昆虫的植物
- 酒店绿墙每块墙板的吸水量是土壤体积的35%~40%，相当于7,500升水
- 两个集水箱，储存雨水，用于灌溉
- 集水箱能存储10,000升雨水
- 绿墙承载了16吨土壤
- 绿墙设备重量6吨
- Treebox公司开发的绿墙设备，由回收利用的聚丙烯材料制成，英国制造
- 植物种类的选择特别注意四季色彩的变化，让外墙在蓝色、粉色、紫色、白色和黄色间变化，草本植物营造出绿色的背景

（表格中的本地植物指是否为英国本土植物，传粉植物指是否为英国皇家园艺学会指定的传粉植物）

ELEVATION

1. Green wall 190m^2
2. Green wall 125m^2
3. Green wall 166m^2
4. LED up-lighting to green walls
5. Water catchment

立面图

1. 绿墙（190平方米）
2. 绿墙（125平方米）
3. 绿墙（166平方米）
4. 绿墙LED照明（向上）
5. 集水区

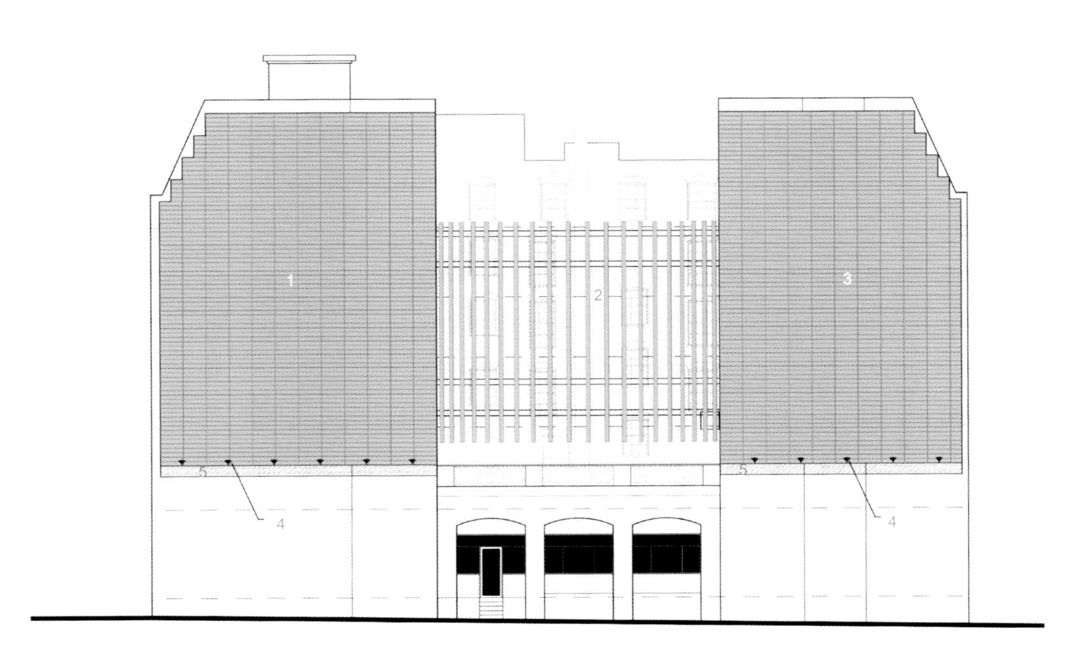

Ode to Burle Marx

布雷・马克斯颂歌

Completion date:
2011
Location:
Mexico City, Mexico
Landscape architect:
VERDE VERTICAL ©
Photographer:
Santigo Arau
Area:
120 sqm

竣工时间：
2011年
项目地点：
墨西哥，墨西哥城
景观设计：
"垂直绿化"设计公司
摄影师：
圣地亚哥・阿劳
面积：
120平方米

Project description:

In the beginning of the 20th century the Brazilian Roberto Burle Marx, one of the most influential landscape architects, started to design his amazing gardens. Initially influenced by the typical European parks and gardens, with its squared geometry, he started to defend the use of the Brazilian flora, studying and collecting it to get a better knowledge about it and then use it in his projects, struggling with the usual overvaluation of exotic plants.

Using the work of plastic artist, and naturalist Burle Marx as a concept, the design of this vertical garden represents one of his inventions in a contemporary graphical way by taking his horizontal plane, and transforming it into a vertical one, allowing us to appreciate it from a different perspective.

The population of cities had grown intensely, the same density forbids the possible existence of parks and green areas, and buildings are still in a construction process. The importance of having nature as close as possible is not only ethical but also an environmental benefit. This is why in the 1960s the possibility of growing plants on a textile substrate and not only on the ground emerged.

In most cities we can find millions of square metres of vertical surfaces that exist with no apparent use just like walls, and can seem abandoned but all of these are potential green surfaces and they can become the structure for the growing need of having nature. The intention of vertical gardens just like this one is to give a response to a need and problem that many cities share which is the growth and population of cities.

20世纪初，巴西著名景观设计师罗伯托·布雷·马克斯（1909－1994）开始了他的设计生涯。马克斯受到欧洲园林风格的影响，大量使用正方形结构，并且强调使用巴西本土植物。他研究并收集巴西植物的品种，积累了丰富的知识，并应用到他的设计中，抵御当时过多使用外来植物的风气。

本案的设计理念就来自园林造型艺术大师布雷·马克斯的作品，把他曾经以现代平面设计手法打造的景观搬到立面上去，变“平面花园”为“垂直花园”，让我们能从一个不同的角度欣赏大师的设计。

当今城市人口不断增长，严重限制了公园和绿地的开发，建筑物还在不断兴建。贴近大自然不仅是我们的精神需求，而且也是环境的要求。这就是为什么20世纪60年代兴起一股热潮，以织物作为基底来培养植物，而不只是让植物在地面生长。

在大多数城市中，都有数百万平方米的垂直表面。这些表面大多没有什么明显的用途，大部分只是墙壁而已，有些甚至看上去有点儿破败，可是这些地方都可以是潜在的绿墙，能够满足我们不断增长的绿化需求。像本案这样的“垂直花园”，目的就是解决许多城市都面临的问题——城市的发展和人口的增长。

CONCEPTUAL PROCESS
设计理念开发过程示意图

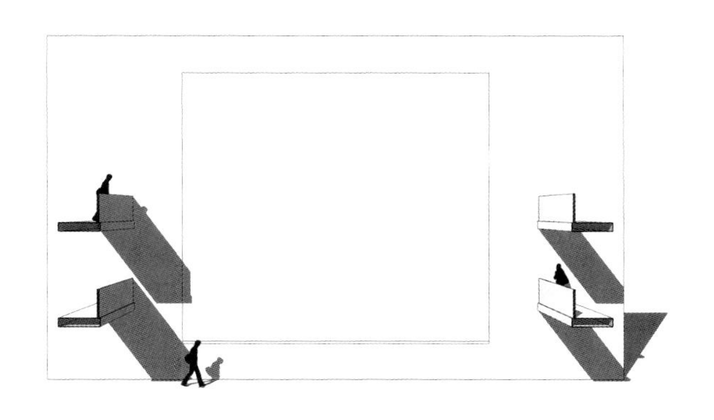
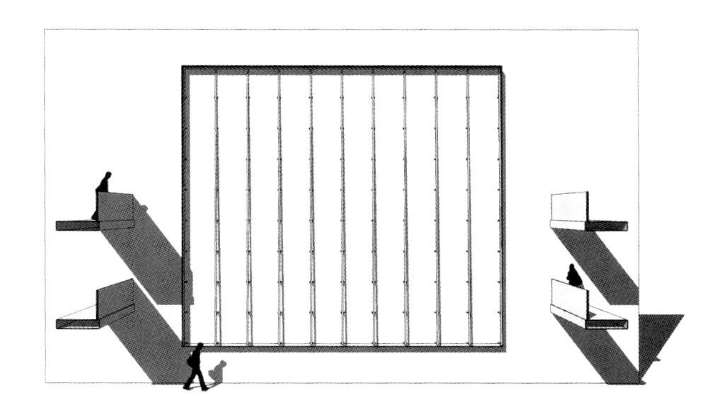
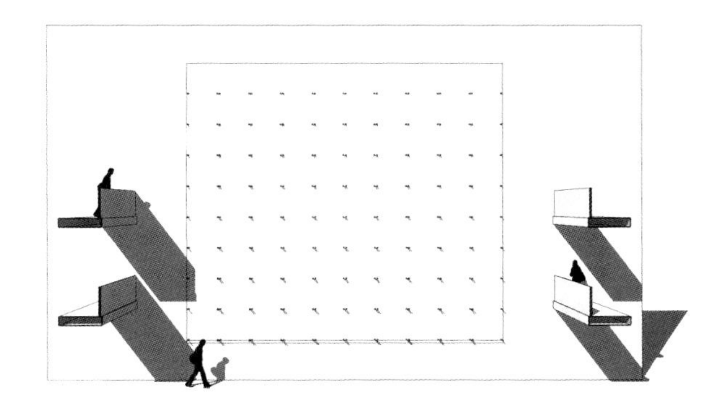
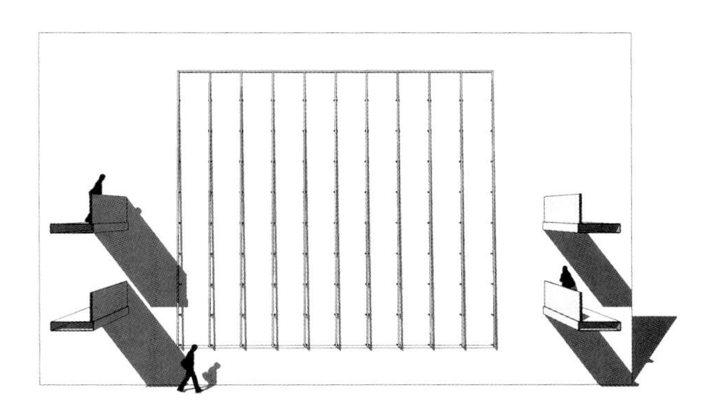
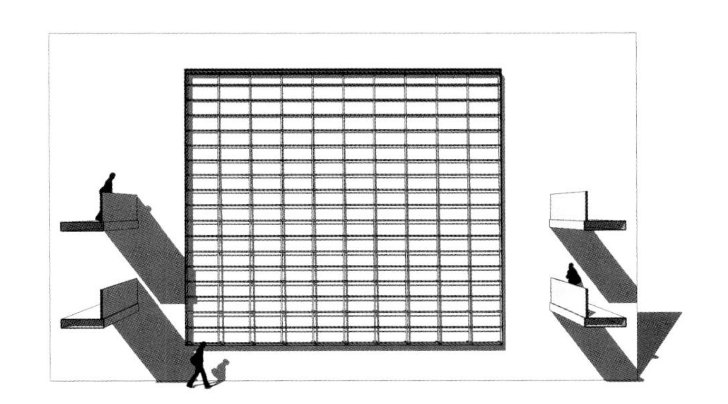
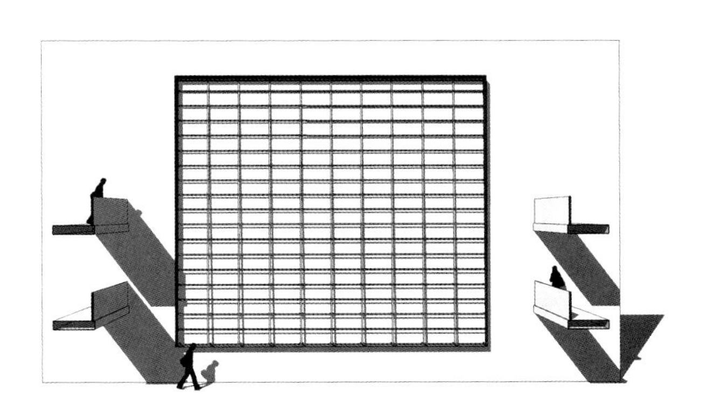

CONSTRUCTION PROCESS 施工过程示意图

Shenzhen Stock Exchange Gardens

深圳证券交易所景观设计

Completion date:
2013
Location:
Shenzhen, China
Architect:
OMA
Landscape architect:
Inside Outside
Green wall consultant:
Verte
Area:
45,000 sqm (1,400 sqm of tropical greens walls, 416 sqm of native living walls)

竣工时间：
2013年
项目地点：
中国，深圳
建筑设计：
大都会建筑事务所
景观设计：
荷兰“内外之间”设计事务所
绿墙顾问：
Verte设计公司
面积：
45,000平方米（热带植被绿墙：1,400平方米；本地植被绿墙：416平方米）

Project description:

The landscape design for the new Shenzhen Stock Exchange consists of four types of gardens that represent and reinforce the new building's architectural and cultural ambition. A spacious park along the north flank of the building and a series of Courtyards, Roof- and Sky Gardens serve as cooling screens for the building's interior and exterior climate and as relaxing areas for the staff.

Since the exchange between the Chinese and Western cultures in the 16th century and the translation of the great Chinese classics into European languages in the 17th century, Confucianism has greatly inspired Western philosophy. This exchange is visualised in the Sino-European gardens where European geometry is combined with the studied asymmetry of the Chinese garden. Elements such as the circle and the path – straight or meandering – are important ingredients in both the Chinese and the Western European garden.

This design considers the Stock Exchange building and its gardens as a place where this mentality of merging cultures and ages could be rendered in a contemporary manner. By interweaving building and garden, interior and exterior, public and private, information and art, aesthetics and function, the design concept aims to integrate the past, present and future.

The East and West Plazas and the North Park encircle the building that stands on a 4.5-hectare site. The granite-covered plazas enter the building's entrance halls and large atrium, passing the Green Screens of the Tropical Garden, until they surrounds the entire base of the tower.

The North Garden is a rectangular public park formed by undulating terraces of waving grasses, into which six mineral, vegetal and aquatic circles lie embedded. A tapestry pattern, based on a European design of the Middle Ages and referring to the traditional Chinese papercut, unfolds into a roof garden on the building's floating podium, offering a multitude of spaces, atmospheres and functions. Inside, the building houses a series of patio- and vertical gardens, spread over its towering height.

Ground Floor Plazas

The general design of the ground floor plazas and park is influenced by the Anglo-Chinese garden style. Each plaza of the ground floor has a distinctive black-grey-white pattern design. A dark diagonal pattern on the East Plaza is laid next to an inverse diagonal layout on the West Plaza and the green patches along the western façade of the tower reflect the same direction and rhythm. The pattern continues inside the atrium and lobby in a more simplified way leaving the light and dark granite forming a checker board. The fact that granite is a local stone makes it ideal for use in the Shenzhen Stock Exchange Square. Based on the heavy use, traffic flow and seasonal humidity, natural stone is best suited to large, public environments. In combination with the almost ethereal white tower, the granite represents the element that stabilises and that connects all areas.

North Park

The North Park consists of circular vegetative and aquatic circles that lie embedded in a continuously undulating terraced plane. The image of a circle inscribed in a rectangle alludes to the harmony in the universe. Several circular gardens escape the rectangular site by enlargement. Three of the six circles are completely filled with trees, contrasting the open meadow. Each one has a specific ground cover and its own tree species. The contour lines of the terraced meadow are lighting up at night, accentuating the topography of the park. Each terrace is filled with a different plant mix: lawn and fields of decorative grasses mixed with perennials, bulbs and herbs that bloom in different seasons. Following the topography, two arc-shaped paths cross the terraced meadow to connect all six circles and to lead the visitor to the various entrances of the building.

深圳证券交易所（SSE）的景观设计由四种类型的花园构成，既突出了这栋新建筑的特点，又体现了中国文化。大楼北侧有一个宽敞的花园，此外还有一系列内庭、屋顶花园和空中花园，对大楼室内外的气候起到屏障的作用，同时也为楼内员工提供了休闲空间。

早在16世纪，中西方文化就有了交流；17世纪，中国的古典名著被翻译成欧洲各国语言传入西方。孔子的儒家思想让西方哲学深受启发。这种文化交流在中西风格相结合的景观设计中体现出来：欧洲的几何造型与中式园林的不对称美相结合，而环形和线形（不论是直线还是弧线）在中西方的园林设计中都是重要元素。

设计师旨在用一种现代的设计手法将深圳证券交易所大楼及其景观打造成东西方文化融合的环境。在这里，建筑与景观、室内与室外、公共与私人、信息与艺术、美观与功能相互交织，设计理念的目标是将过去、现在和将来融为一体。

本案占地面积4.5公顷，东侧和西侧广场以及北侧花园将建筑围在中间。广场的地面铺装采用花岗岩，一直延伸到大楼的入口大厅和中庭里，让“热带花园”的景致深入室内，室内空间环绕在绿色屏障的包围中。

北侧花园的平面布局呈矩形，景观设计采用绵延起伏的草坪，草坪上嵌有6个圆形花坛和水池。裙楼上方的屋顶花园，景观设计呈现出中世纪欧洲的挂毯图案，也颇有中国传统剪纸艺术的神韵。屋顶花园内分出不同的小空间，氛围和功能各不相同。大楼内部也进行了绿化，不但规划出一系列景观内庭，还打造了室内绿墙，分布在楼内不同的高度上。

一楼广场

一楼广场和花园的设计对中西园林设计风格都有体现。一楼每个广场的设计都采用特别的“黑白灰”三色组合。东侧广场采用深色的斜线图案，西侧广场采用相反方向的斜线。大楼西侧的点点绿色也呈现出相同的方向和韵律。这种图案设计一直延伸到室内，中庭和大厅里用深色和浅色的花岗岩拼成棋盘状的图案。由于花岗岩是中国常见的石材，所以用于深圳证券交易所广场的地面铺装最理想不过了。由于使用非常频繁，车辆人流川流不息，再加上季节性的湿度变化，所以天然石材是大型公共环境中最合适的铺装材料。深圳证券交易所的白色大楼高耸入云，轻盈飘渺，下面的花岗岩铺装将其牢牢固定在地面上，并且花岗岩也是各个区域的连接元素。

北侧花园

北侧花园地势起伏不平，种植草坪，里面嵌有圆形花池或水生植物池。矩形中间镶嵌圆形，这样的图案喻指宇宙的和谐。圆形花池跳脱出矩形场地的刻板限制。6个花池中的3个种植的全是树木，跟周围的开放式草坪形成对照。每个花池的地面都有特殊的铺装，树木种类也各不相同。草坪根据地势高低分为几块，每一块的边缘都进行夜间照明，进一步凸显花园的地势。每一块草坪上面种植的植物也有不同，除了平常的草本植物外，还有观赏性植物，搭配多年生植物、鳞茎类植物和药草类植物，在不同的季节开花。两条弧形小径沿着起伏的地势穿过各个草坪，将6个花池连接起来，将访客引至大楼的各个入口。

Podium Garden

On a tremendous overhanging roof, floating at a height of 60 metres above street level, a roof garden was created for the extensive staff of the Shenzhen Stock Exchange firm. The floral mosaic of the Podium Garden is based on a European pattern designed in the Middle Ages. Yet the pattern could also be seen as a Chinese paper cut-out. This double interpretation is part of the landscape concept idea in which the merging of cultures is addressed.

屋顶花园

裙楼上方的屋顶非常宽敞，距离地面标高60米，仿佛一座飘浮在空中的花园，供深圳证券交易所的员工休闲使用。屋顶花园的植被设计以中世纪欧洲的景观形制为基础。同时，中国的剪纸艺术也在其中有所体现。这种中西合璧的手法是整体景观设计理念的一部分，强调两种文化的融合。

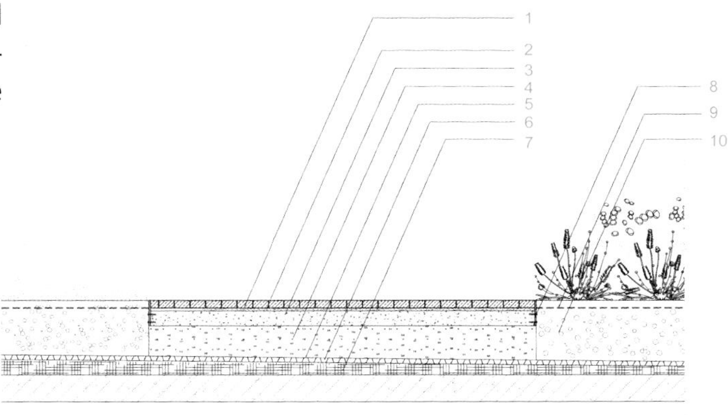

PLANT EGDE DETAIL1

1. Granite tiles, 100x100x50 thk hammered
2. 20mm sand
3. Concrete
4. Lightweight filling lava stones
5. Drainage and water retention mat
6. Root control
7. Waterproof layer
8. Planted edge steel
9. Drip irrigation
10. Soil mix with expanded clayballs

种植边缘详图-1

1. 手凿面花岗岩（100毫米×100毫米×50毫米）
2. 砂浆层（20毫米厚）
3. 混凝土
4. 轻质填充（火山岩）
5. 排水保水层
6. 阻根层
7. 防水层
8. 种植边缘（不锈钢）
9. 滴灌管道
10. 土壤混合膨胀土球

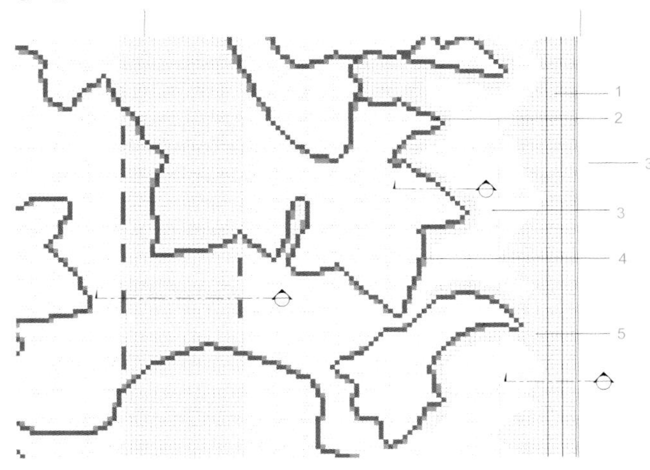

PODIUM PARTIAL PLAN

1. Drainage gutter
2. Shanxi granite flamed 100x100x50 thk
3. Planting
4. Granite hammered 100x100x50 thk
5. Granite hammered 100x100x50thk

屋顶花园平面图

1. 排水槽
2. 山西火烧面花岗岩（100毫米×100毫米×50毫米）
3. 绿化种植
4. 手凿面花岗岩（100毫米×100毫米×50毫米）
5. 手凿面花岗岩（100毫米×100毫米×50毫米）

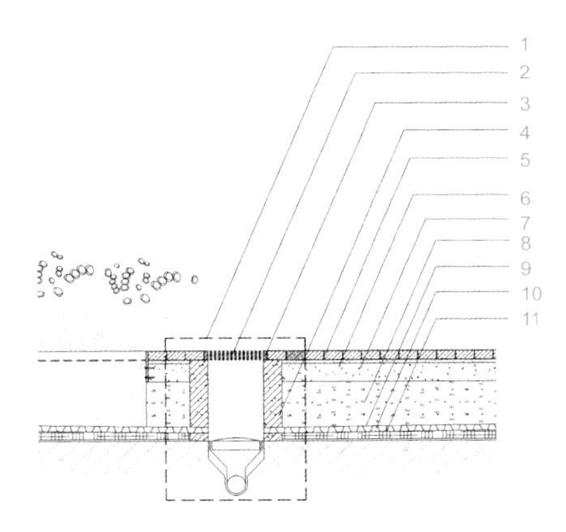

PLANT EGDE DETAIL2

1. Detail
2. Stainless steel grating
3. Drainage gutter
4. Concrete
5. Granite tiles, 100x100x50 thk, hammered
6. 50mm sand
7. Filted mat
8. Lightweight filling lava stones
9. Drainage and water retention mat
10. Root control
11. Waterproof layer

种植边缘详图-2

1. 细部
2. 不锈钢水篦子
3. 排水槽
4. 混凝土
5. 手凿面花岗岩（100毫米×100毫米×50毫米）
6. 砂浆层（50毫米厚）
7. 过滤层
8. 轻质填充（火山岩）
9. 排水保水层
10. 阻根层
11. 防水层

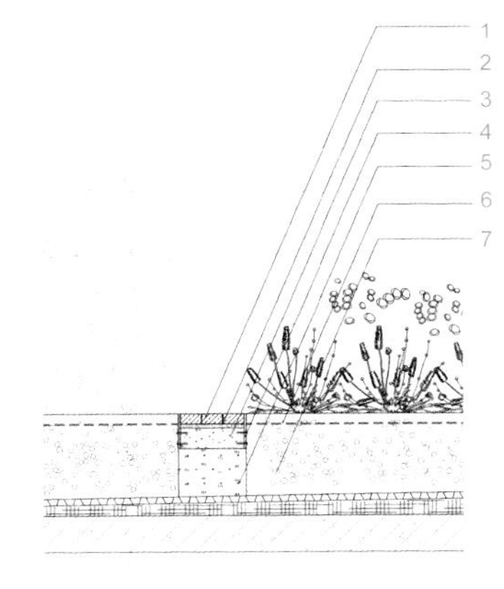

PLANT EGDE DETAIL3

1. Granite tiles, 100x100x50 thk, hammered
2. 20mm sand
3. Concrete
4. Planted edge steel
5. Lightweight filling lava stones
6. Drip irrigation
7. Soil mix with expanded clayballs

种植边缘详图-3

1. 手凿面花岗岩（100毫米×100毫米×50毫米）
2. 砂浆层（20毫米厚）
3. 混凝土
4. 种植边缘（不锈钢）
5. 轻质填充（火山岩）
6. 滴灌管道
7. 土壤混合膨胀土球

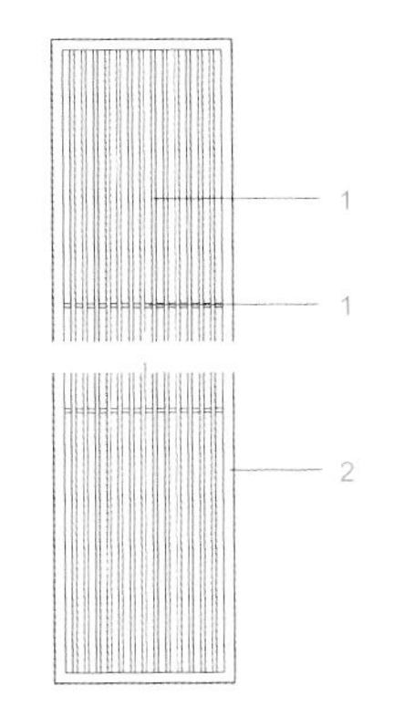

GRATING DETAIL

1. 8x50 thk stainless steel polished finish
2. 20x50 thk stainless steel polished finish

水篦子大样图

1. 不锈钢条（8毫米×50毫米）
2. 不锈钢条（20毫米×50毫米）

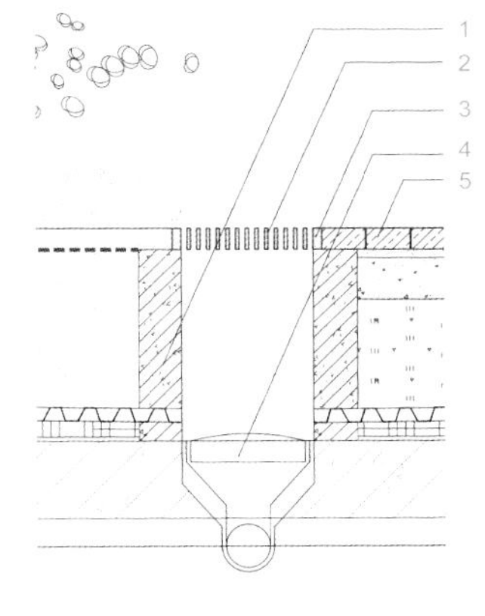

GRATING DETAIL

1. Concrete
2. 8x50 thk stainless steel polished finish
3. 0x50 thk stainless steel polished finish
4. Drainage gutter
5. Granite tiles, 100x100x50 thk, hammered

水篦子大样图

1. 混凝土
2. 不锈钢条（8毫米×50毫米）
3. 不锈钢条（20毫米×50毫米）
4. 排水槽
5. 手凿面花岗岩（100毫米×100毫米×50毫米）

This green "carpet" is formed by filling in the line drawing (measuring 161x 97 metres) with trimmed-up shrubs and ground-cover plants that form screens of different heights; or soft, inviting carpets. To achieve a variation of spaces and spatial effects, five different planting types are used: high shrubs, low shrubs, perennials, ornamental grasses and lawn.

Each plant has its own role to fulfill; the high shrubs (2m) protect from wind and sun and the low shrubs (1m) define areas. The grasses and flower bulbs provide colour, movement and light. The perennials offer changing colours and scents. The lawn tempers the acoustic and gives a soft feel to the roof garden's surface.

屋顶的"绿毯"大小为161米x 97米，仿佛一张线条画，线条处是修剪整齐的灌木和地表植物，灌木形成不同高度的绿色屏障，地表植物形成一块块柔软的"绿毯"。为了实现空间的多样性，设计师采用了5种不同的植物类型：较高灌木、低矮灌木、多年生草本植物、观赏性禾本植物和草坪。

每种植物有自己的功能。较高的灌木（2米高）起到保护作用，抵御风吹日晒；低矮灌木（1米高）起到界定空间的作用。禾本植物和开花的鳞茎类植物带来色彩和动感。多年生草本植物营造色彩和芳香的变化。草坪起到吸收噪声的作用，赋予屋顶花园以柔软的质感。

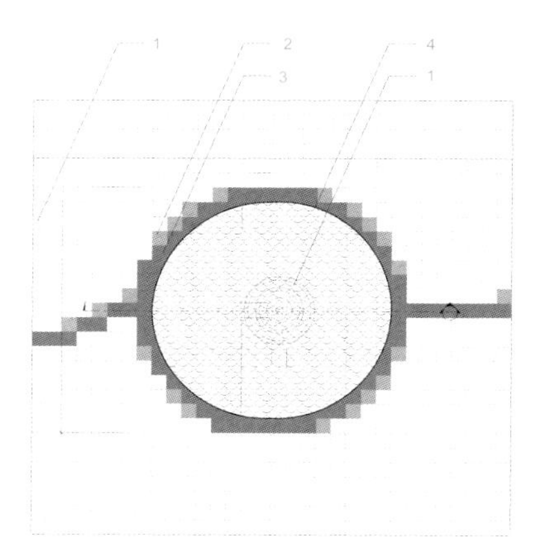

POND PLAN

1. Granite hammered 100x100x50 thk
2. Granite hammered 100x100x50 thk
3. Shanxi granite flamed 100x100x50 thk
4. Stainless steel

水景平面图

1. 手凿面花岗岩（100毫米×100毫米×50毫米）
2. 手凿面花岗岩（100毫米×100毫米×50毫米）
3. 山西黑色火烧面花岗岩（100毫米×100毫米×50毫米）
4. 不锈钢

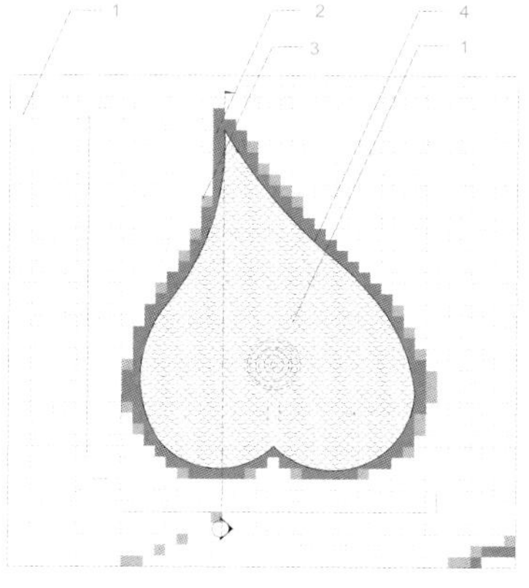

POND PLAN

1. Granite hammered 100x100x50 thk
2. Shanxi granite flamed 100x100x50 thk
3. Granite hammered 100x100x50 thk
4. Stainless steel

水景平面图

1. 手凿面花岗岩（100毫米×100毫米×50毫米）
2. 山西黑色火烧面花岗岩（100毫米×100毫米×50毫米）
3. 手凿面花岗岩（100毫米×100毫米×50毫米）
4. 不锈钢

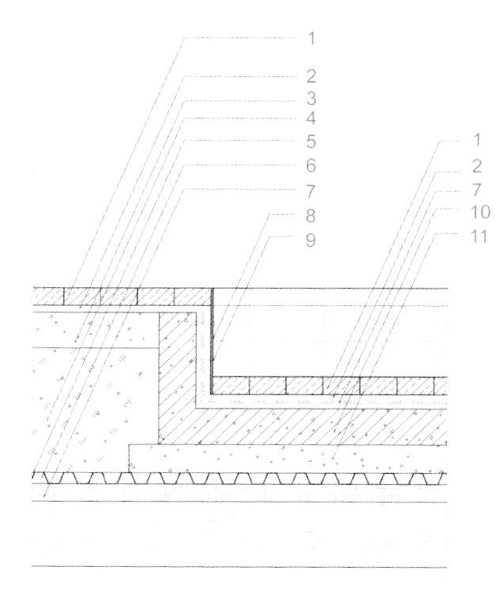

POND DETAIL

1. Granite hammered 100x100x50 thk
2. 50mm sand
3. Filted mat
4. Lightweight filling lava stones
5. Drainage and water retention mat
6. Root control
7. Waterproof layer
8. Stainless steel
9. Bolt
10. Ferroconcrete
11. Concrete

水景大样图

1. 手凿面花岗岩（100毫米×100毫米×50毫米）
2. 砂浆层（50毫米厚）
3. 过滤层
4. 轻质填充（火山岩）
5. 排水保护层
6. 阻根层
7. 防水层
8. 不锈钢
9. 螺栓
10. 钢筋混凝土
11. 混凝土

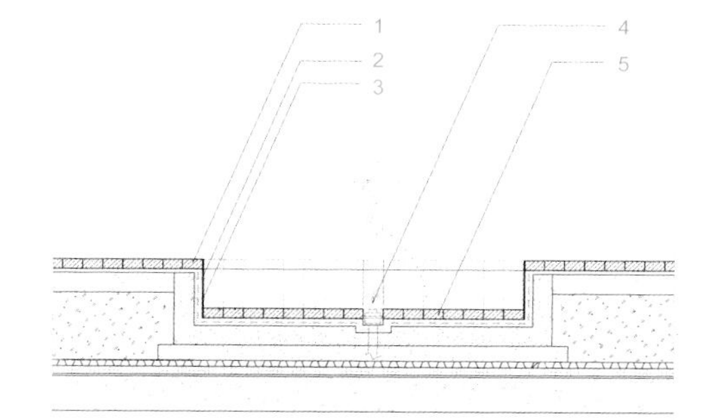

POND SECTION

1. Shanxi granite flamed 100x100x50 thk
2. Stainless steel
3. Bolt
4. Fountain device
5. G633 granite hammered 100x100x50 thk

水景剖面图

1. 山西黑色火烧面花岗岩（100毫米×100毫米×50毫米）
2. 不锈钢
3. 螺栓
4. 喷头
5. G633浅色手凿面花岗岩（100毫米×100毫米×50毫米）

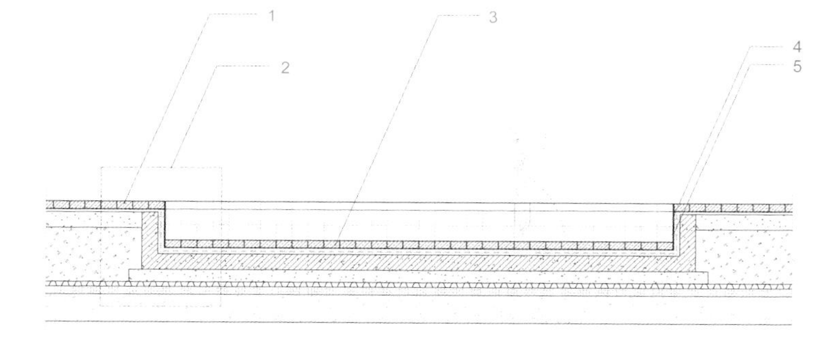

POND SECTION

1. Shanxi granite flamed 100x100x50 thk
2. Detail
3. G633 granite hammered 100x100x50 thk
4. Stainless steel
5. Bolt

水景剖面图

1. 山西黑色火烧面花岗岩（100毫米×100毫米×50毫米）
2. 细部
3. G633浅色手凿面花岗岩（100毫米×100毫米×50毫米）
4. 不锈钢
5. 螺栓

Together, these five ingredients form a composition of screens and surfaces that differ in height and thickness (between transparent and opaque), in structure and colour; and in seasonal effect (evergreen, seasonal, flowering, fruit- or seed bearing). The pattern organises the given space in a functional manner and invites for different uses including meeting, strolling, looking out, relaxing, sitting and jogging.

The local granites proposed on the Ground Floor and First Floor Plazas are continued on the Podium surface. Treated with the same finishes – hammered and flamed – the tiles will have different tones of grey; from very light (almost white) to very dark (almost black). The marking of a path around the building which can as well be used for running is made out of granite with a hammered finish to avoid slipperiness.

Courtyards

Four double high courtyards with a surface area of 10.4 x 4.4 metres give a green atmosphere to the offices on the 9th floor. As these Courtyards are seen as lowered parts of the Podium Garden, the planting pattern of the podium is continued here. Each courtyard is enclosed with glass and all are completely visible from the interiors. Spiral staircases make the courtyards accessible from the podium level. They will host various climbing species which will add verticality to these spaces and are well known for their scent and colour: Jasmine, rose, white wisteria, and clematis. Echoing the species selected for the Podium Garden, the courtyards will be embellished with shrubs and perennials and offer terrace space for the staff to relax.

The 10x10 cm light and dark granite tiles will also be the applied floor finish, as well as the furniture that is proposed for the areas on the podium above.

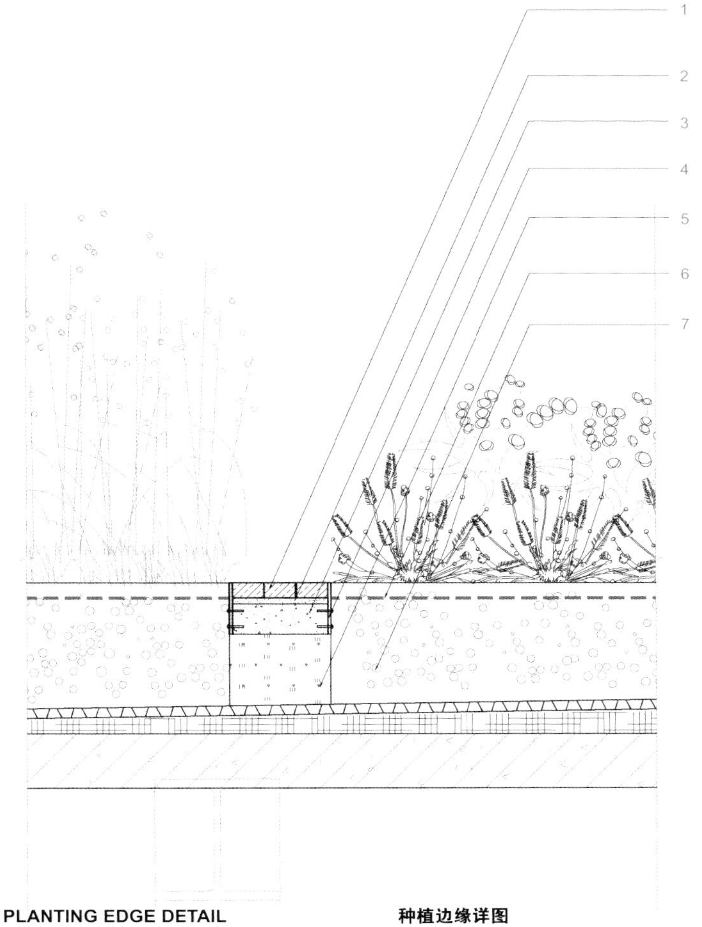

PLANTING EDGE DETAIL

1. Granite tiles, 100x100x50 thk, hammered
2. 20mm sand
3. Concrete
4. Planted edge – steel
5. Lightweight filling – lava stones
6. Drip irrigation
7. Soil mix with expanded clayballs

种植边缘详图

1. 手凿面花岗岩（100毫米×100毫米×50毫米）
2. 砂浆层（20毫米厚）
3. 混凝土
4. 种植边缘（不锈钢）
5. 轻质填充（火山岩）
6. 滴流灌溉
7. 土壤混合膨胀土球

这5种植物共同组成绿色“屏风”和“地毯”，高度、厚度各有不同（有些地方透明，有些地方半透明），造型、色彩也有差异，并且将在不同的季节呈现出不同的效果（有些是常绿的，有些随季节荣枯、开花、结果或者结籽）。

屋顶花园的植被布局界定出一系列不同功能的空间，能够满足不同的使用需求，如聚会、散步、观景、休息、闲坐或慢跑。

一楼和二楼的广场采用的当地花岗岩石材，在屋顶花园上也延续使用，并且采用相同的工艺手法（捶纹、火烧）。花岗岩表面呈现出不同色调的灰色，从非常浅的灰（接近于白色）到深灰（接近于黑色）。沿建筑周围设置一圈小径（也可以用作跑步道），小径的铺装采用防滑的捶纹花岗岩。

内庭

4个双层通高的内庭（大小为10.4 米x 4.4米）为9楼的办公区带来一丝绿意。从这些内庭可以看到下方裙楼上的屋顶花园，所以屋顶花园上的植被形态在内庭中也延续使用。每个内庭都用玻璃隔墙围合起来，从室内就能看到庭院内部。几部旋转楼梯从裙楼直通内庭。楼梯上种植了各种攀援植物，打造垂直景观，而且这些植物都以其色彩和香味著称，包括茉莉、玫瑰、白色紫藤和铁线莲等。内庭的植被种类也与屋顶花园的相同，以灌木和多年生草本植物为主，并且辟出一个平台，供员工休息之用。

内庭地面的花岗岩石板颜色深浅不一，尺寸为10厘米见方，地面铺装和家具小品都与屋顶花园相同。

LOW SHRUBS 低矮灌木

Carmona microphylla 基及树

Buxus semprervirens 锦熟黄杨

Gendarussa vulgaris 小驳骨

Pittosporum tobira 海桐

Aglaia duperreana 四季米仔兰

Duranta repens 假连翘

Murraya exotica 九里香

Jasminum sambec 茉莉花

HIGH SHRUBS 高大灌木

Scheffl era arboricola 鹅掌藤

Olea ferruginea 锈鳞木犀榄

Viburnum odoratissimum 珊瑚树

Taxus baccata 红豆杉

GRASSES + BULBS 草本 + 球茎

Narcissus triandrus 'Thalia'
水仙

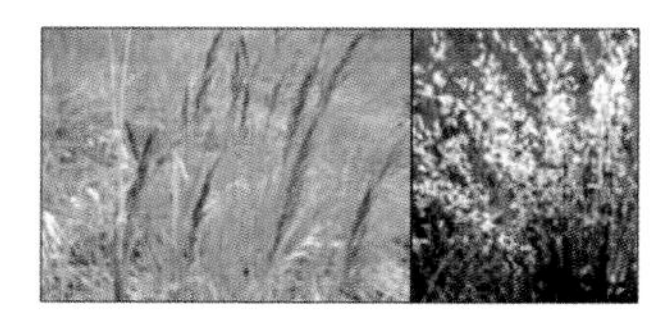

Andropogon scoparius
小须芒草

Penisetum setaceum
紫叶狼尾草

Calamagrostis brachytricha
宽叶拂子茅

Miscanthus sinensis 'Ghana'
芒草

Miscanthus fl oridulu
五节芒

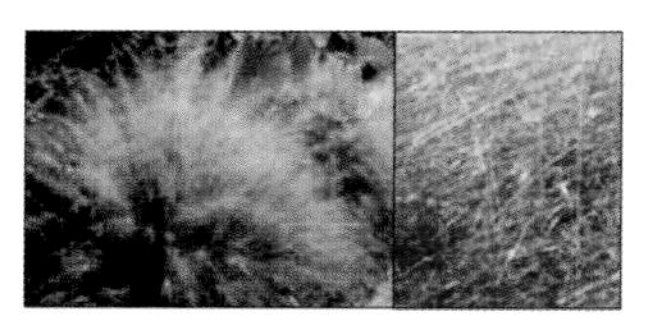

Muhlenbergia capillaris
毛芒乱子草

Allium schoenoprasum
香葱

PERENNIALS 多年生植物

Crocosmia lucifera
火星花

Lavandula angustiolia
薰衣草

Liriope spicata
山麦冬

Melastoma candidum
野牡丹

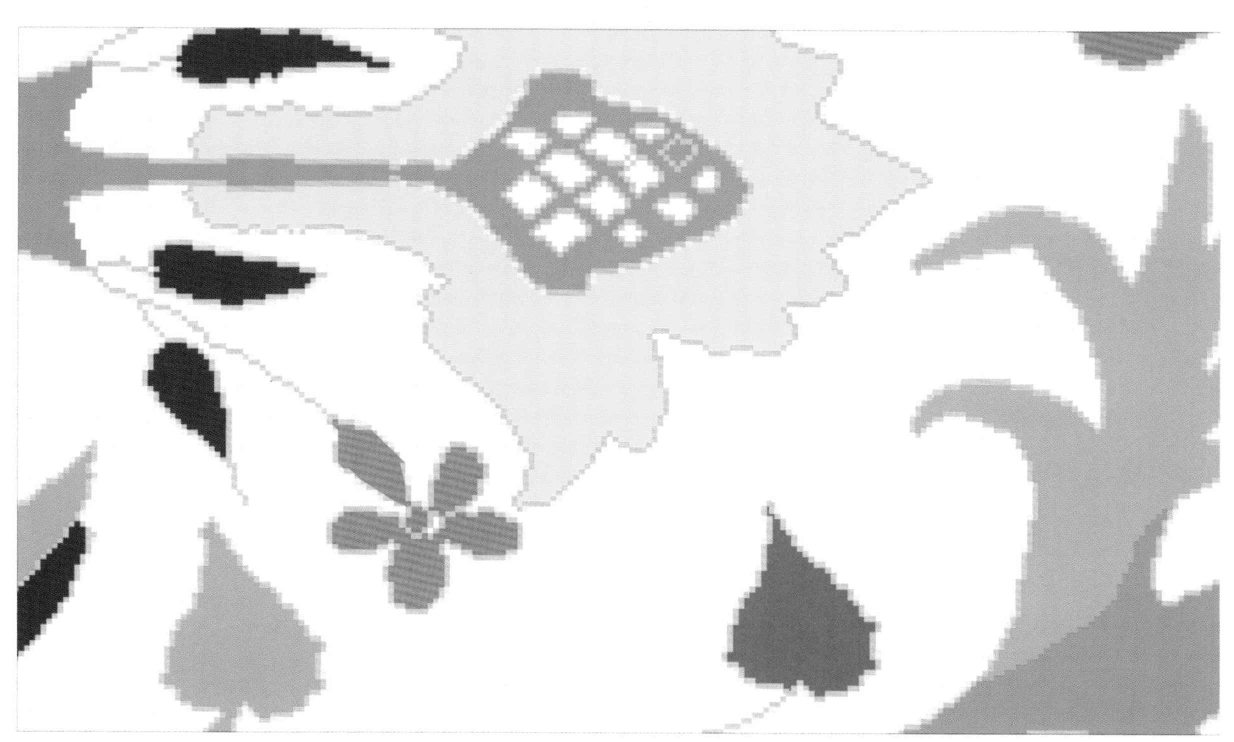

PLANTING PLAN 植物分布图

Ixora chinensis
龙船花

Hemerocallis fulva
萱草

Nicotina sanderae
花烟草

Gompherena globosa amaranthaceae
千日红

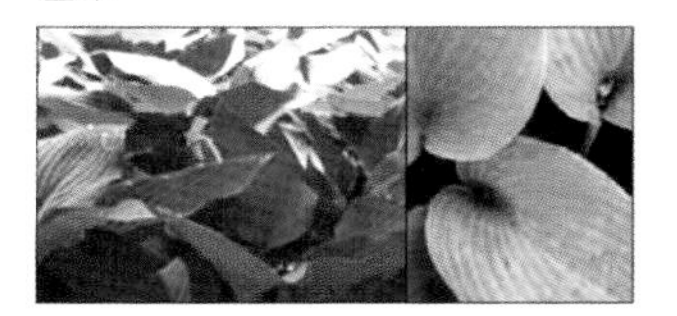

Hosta 'Halcyon'
翠绿玉簪

Lycoris radiate
石蒜

Cleome hassleriana
醉蝶花

Salvia coccinea
蜂鸟鼠尾草

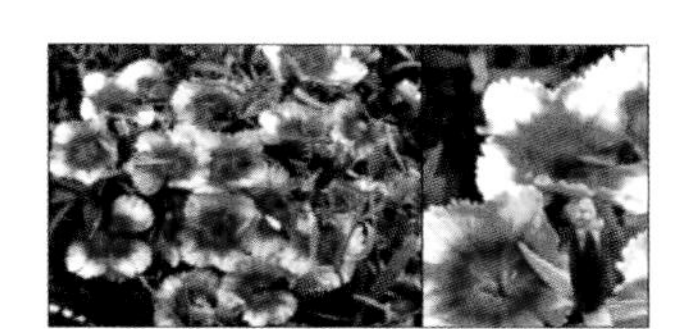

Dianthus chinensis
石竹

Myosotis hortensis
勿忘草

Ageratum conyzoides
藿香蓟

Mentha spicata
留兰香

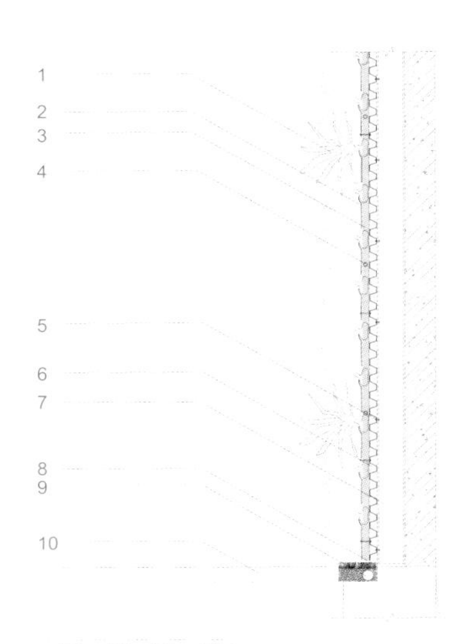

COOLING TOWERS DETAIL

1. Everygreen ferns
2. 10mm felt + plastic layer with slit as pocket
3. Second 50mm felt layer
4. Hose for irrigation
5. Bolt
6. 50mm metal sheeting support structure (EG stainless steel)
7. 180mm I section profile (EG stainless steel)
8. Drainage gutter covered by granite gravel
9. Drain pipe
10. Granite tiles

冷却塔细部大样图

1. 常绿蕨类植物
2. 毡子和塑料层，10毫米，有割缝，用于种植
3. 第二层毡子，50毫米
4. 灌溉软管
5. 螺栓
6. 不锈钢板支撑结构，50毫米
7. 工字型不锈钢，180毫米
8. 排水槽（铺设一层花岗岩碎石）
9. 排水管
10. 花岗岩砖

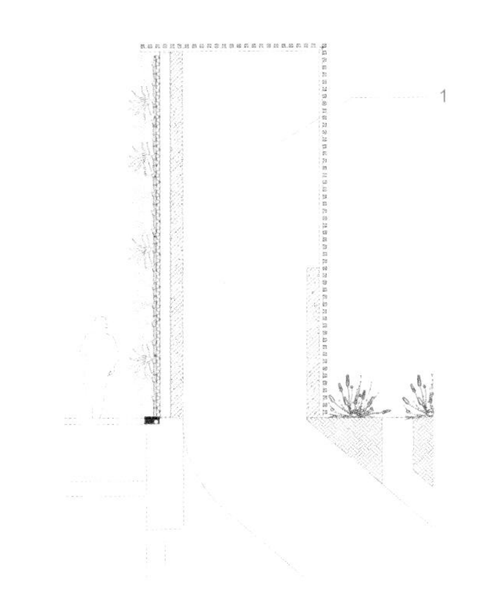

COOLING TOWERS SECTION

1. Cooling tower

冷却塔剖面图

1. 冷却塔

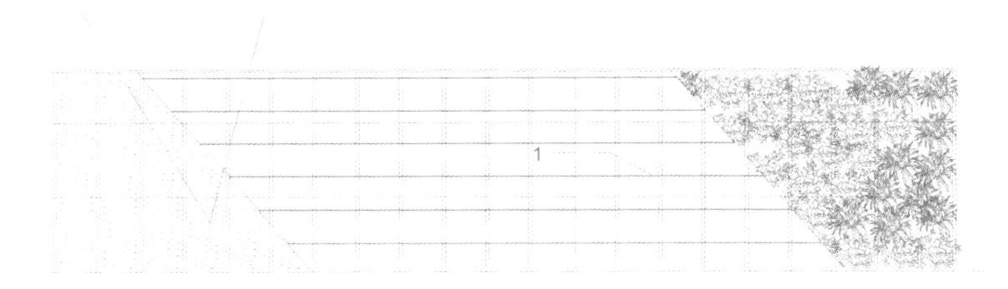

POCKET SLIT DETAIL ZOOM

1. Drip irrigation pipe

割缝大样图

1. 滴流灌溉管

COOLING TOWERS SOUTH ELEVATION

1. Stainless pipe 100x100 thk
2. Stainless pipe 50x25x1.6 thk
3. Fabric mesh
4. Different plant

冷却塔南侧立面图

1. 暗灰色不锈钢方通（100毫米 × 100毫米）
2. 暗灰色不锈钢方通（50毫米 × 25毫米 × 1.6毫米）
3. 织物网格
4. 不同植物

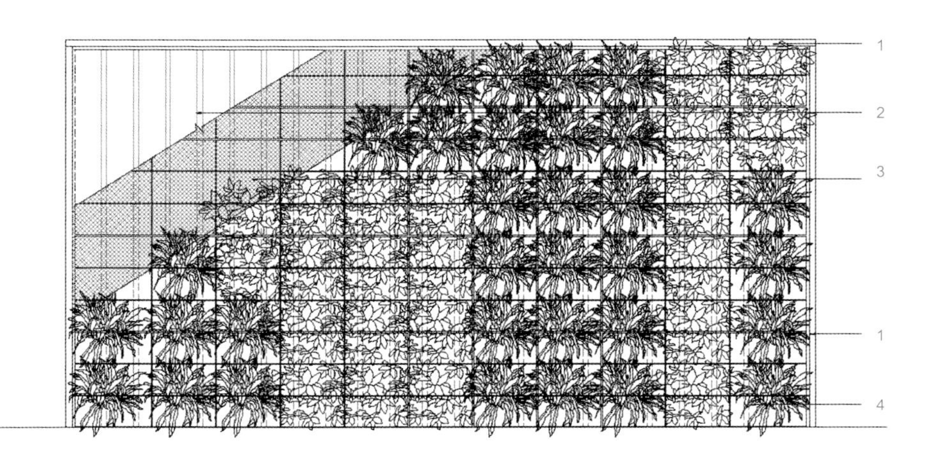

SOUTH PLANTING ELEVATION OF COOLING TOWER
1. Groundcover
2. Fern
3. Creeper

冷却塔南侧植物立面图
1. 种植地被植物
2. 种植蕨类植物
3. 种植藤蔓植物

PLANTING SECTION OF COOLING TOWER
1. Fern
2. Creeper
3. Terraced garden

冷却塔植物剖面
1. 种植槽内种植蕨类植物
2. 种植槽内种植藤蔓植物
3. 阶梯花园

BOTANICAL NAME 植物学名	HEIGHT(CM) 自然高度（厘米）	AREA(M^2) 面积（平方米）	COMMENT 备注
Tripogandra cordifolia 怡心草	5-10	240	Everygreen, fine and legant 生长迅速，叶色清雅明丽
Carex follosiss 白玲苔	15-20		Everygreen, light green leaves 叶色清雅
Scirpus cernuus 孔雀蔺	15-20		Everygreen, fine and elegant, comeliness and standing 四季常青，纤细优雅，清秀挺拔
Dichondra micrantha 马蹄金	5-10		Dense leaf, evergreen 叶片密集，四季常青
Calathea lubbersiana 黄斑竹芋	30-40	200	Elegant planting, yellow strip on the leaf; prefer half-shady environment with warm and moist 株态秀雅，叶具精致的黄色斑纹；喜温暖、湿润的半阴环境
Calathea zebrine 绒叶肖竹芋	30-40		Elegant planting, zebra strip on leaf, velour luster; prefer half shady environment with warm and moist 株态秀雅，叶具斑马状条纹，天鹅绒光泽；喜温暖、湿润的半阴环境
Calathea rotundifolia 圆叶竹芋	30-40		Elegant planting, mellow and unique leaf form; prefer half-shady environment with warm and moist 株态秀雅，叶形圆润独特；喜温暖、湿润的半阴环境
Calathea bicolor 双色竹芋	25-40		Elegant planting, beautiful leaf pattern; prefer half-shady environment with warm and moist 株态秀雅，叶片花纹美丽；喜温暖、湿润的半阴环境
Calathea setosa 银羽竹芋	30-40		Elegant planting, silver, yellow, green feather leaf; prefer half-shady environment with warm and moist 株态秀雅，有银、黄、绿色羽毛般的条纹；喜温暖、湿润的半阴环境
Passionfora edulia 西番莲	25-30	50	Beautiful planting, unique leaf form, white flower with a little bit purple, purple fruits 植株美观，叶形别致；花白色稍带紫晕；果实圆润，紫色
Lonicera japonica 金银花	25-30		Beautiful planting, flower first as silver white, turn yellow after a couple of days, delicate fragrance 株型美观，花初放时洁白如银，数天后变成黄色，清香随风四溢

Cooling Towers in the North Park

Seven cooling towers in the North Park together create a tall, almost continuous wall. The designers transformed the backside of the structures into an attractive front by planting the 6.20-metre-high walls with native forest plants. Those evergreen plants create lush vertical landscapes and make it a true “living wall” that aligns the public path along the building’s north façade. Seven screens of each 9.60m add up to a total green surface of 416 square metres!

北侧花园的冷却塔

北侧花园中有7座冷却塔，几乎形成一面高墙。设计师在冷却塔的后墙上栽种了6.20米高的本地植物，茂盛生长的植物将背面变成了正面。这些常绿植物营造出丰富的垂直景观，形成一面真正的“活的绿墙”，让大楼北侧的道路通行变得多姿多彩。这7面“绿色屏风”（每面宽9.60米）共带来416平方米的绿化面积。

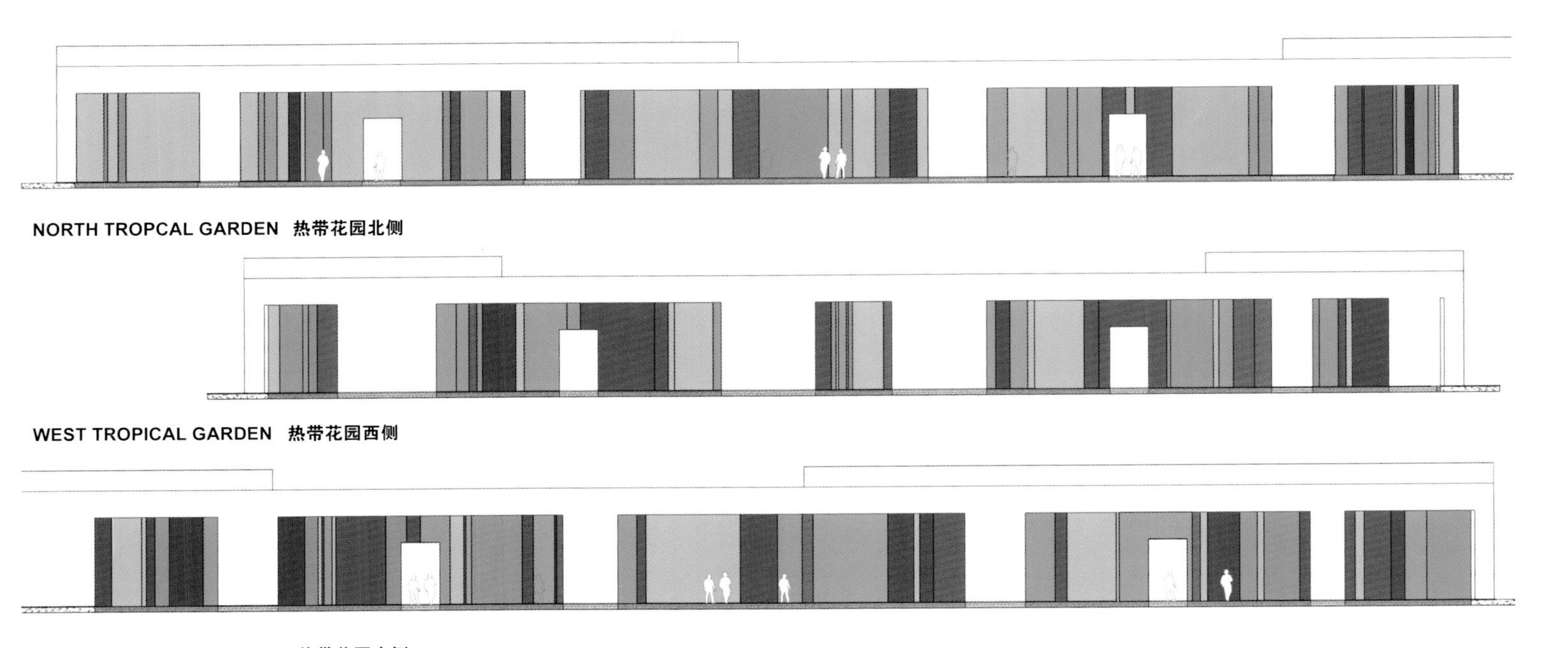

NORTH TROPCAL GARDEN 热带花园北侧

WEST TROPICAL GARDEN 热带花园西侧

SOUTH TROPICAL GARDEN 热带花园南侧

TROPICAL GARDEN PLAN 热带花园平面

1. Green wall 1. 绿墙
2. Entrance 2. 出入口
3. Water channel 3. 水渠
4. Bridge 4. 人行天桥
5. White pebbles 5. 白色卵石

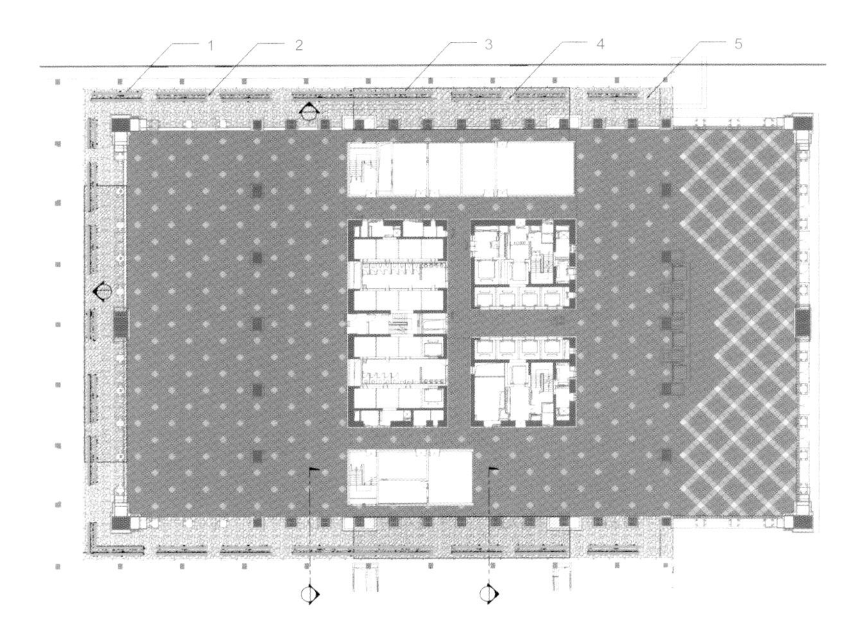

GREEN WALL SECTION 绿墙剖面图

1. U profile stainless steel (5mm) 1. 5毫米厚U形不锈钢片
2. Different plant 2. 植物
3. L-section steel profile 3. 角钢
4. Fabric mesh 4. 固定结构网
5. Amino plastic foam 5. 化学塑料泡沫种植棉
6. Hose for irrigation 6. 灌溉管
7. Moisture sensor 7. 湿度感应器
8. 20mm stainless steel vertical battens 8. 20毫米厚不锈钢垂直板条
9. I-section steel profile 9. 工字钢
10. Pump with water level sensor 10. 水泵（带水量感应器）
11. U-profile stainless steel (5mm) 11. 5毫米厚U形不锈钢片
12. Water 12. 水
13. White pebbles in white cement 100mm 13. 100毫米厚白色水泥嵌白色卵石

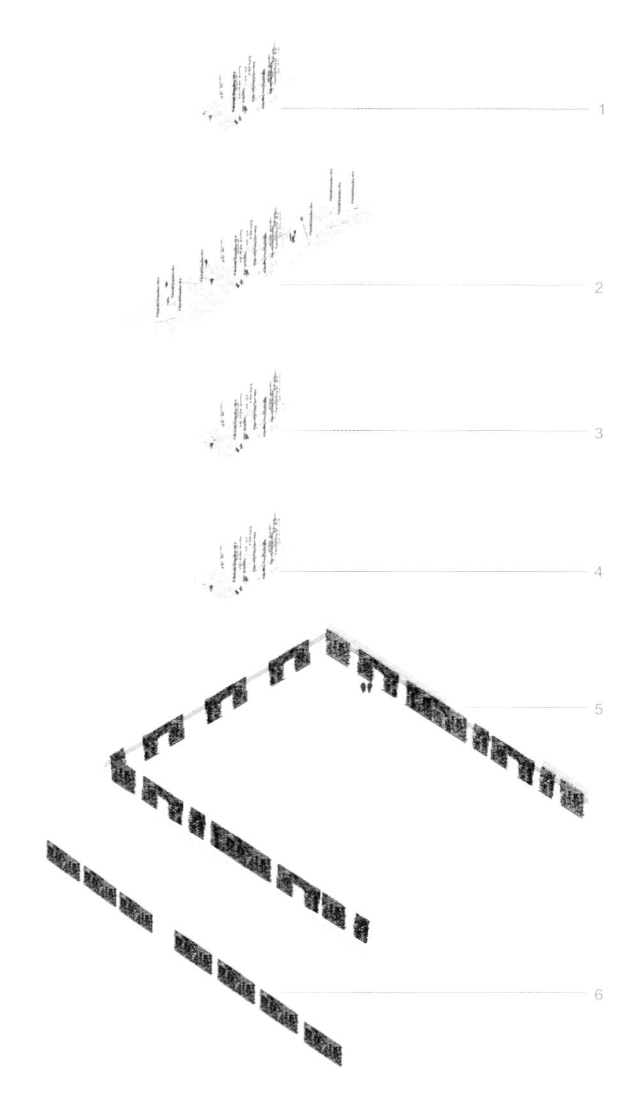

VERTICAL GREEN SCHEME
1. Sky Garden 44F
2. Sky Garden 33F
3. Sky Garden 22F
4. Sky Garden 11F
5. Tropical Garden
6. Green Wall Cooling Towers

垂直绿化示意图
1. 44层空中花园
2. 33层空中花园
3. 22层空中花园
4. 11层空中花园
5. 热带花园
6. 绿墙冷却塔

Tropical Garden

A series of double faced, hydroponic green walls and gates with an impressive height of 4.8 metres and a total length of almost 200m transform this open corridor, encircling the building, into a tropical garden. Bromeliads, ferns and orchids grow over the nutrient wall creating a lush and colourful atmosphere. The plants are all organised in vertical stripes from top to bottom creating an abstract, vegetal painting. The total surface of the green adds up to almost 1,400 square metres.

“热带花园”

“热带花园”是一条绿色走廊，两边都是绿墙，栽种水培植物。绿墙高达4.8米，长度共计200米，围绕着整座建筑，将大楼环绕在热带花园的氛围中。墙上种植的是凤梨科植物、蕨类植物和兰科植物，用营养液栽培，生长非常茂盛，营造出多姿多彩的环境。植物从上到下栽种在垂直种植槽里，营造出一幅抽象的“植被画”。绿色墙面的面积约为1,400平方米。

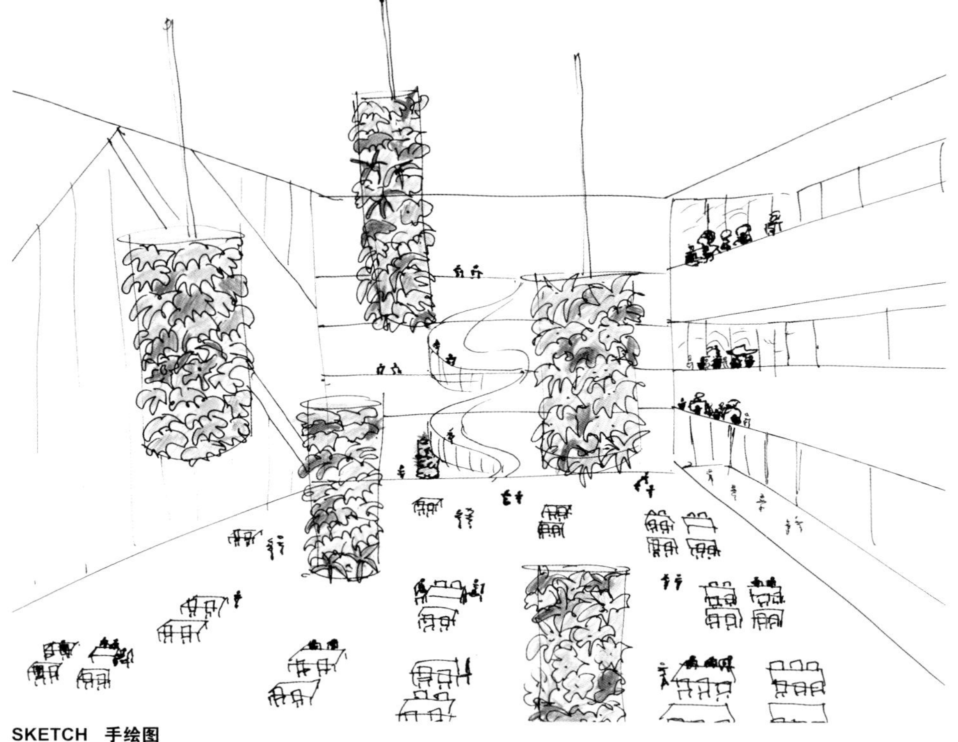

SKETCH 手绘图

Sky Gardens

The sky gardens are high rectangular spaces of 8.4 x 18 metres with a height of 11.6 metres. The sky gardens connect three consecutive floors and have a wide view towards the western side of the SSE plot, the city and the sky. Their interiors are faintly visible from the street and they are to be enjoyed from the interior spaces surrounding them.

Four vertically organised gardens composed of linear elements with a considerable dimension are created. The visitor of the sky gardens will be able to take a closer look and admire the vegetation in detail. The sky gardens situated on the 11th, 23rd, 33rd and 45th floor will each have its own species so they are differentiated by means of colour, texture and scent.

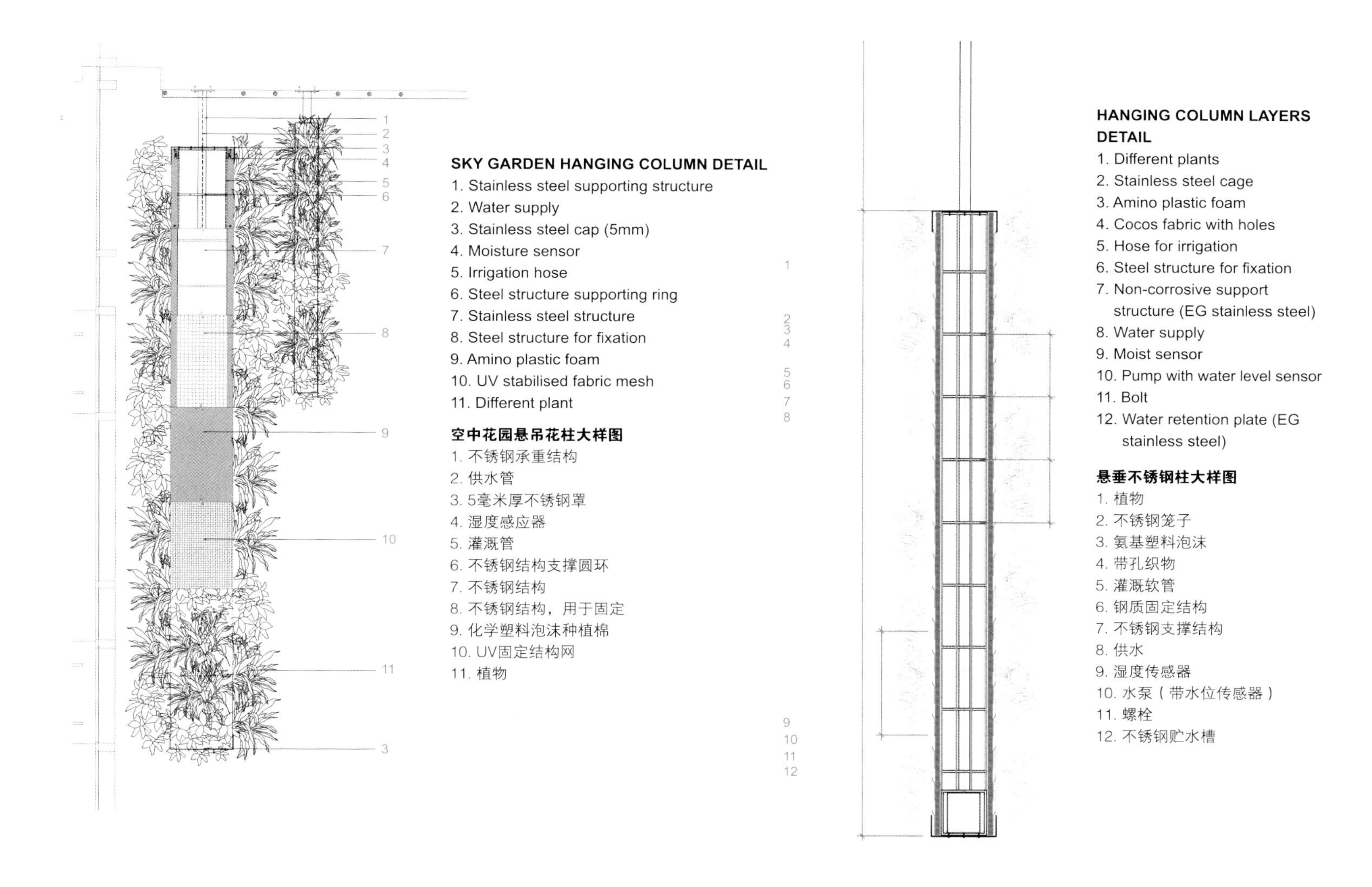

The Tropical and Cooling Tower Gardens of the ground floor – both vegetal walls – reappear as Column Gardens in the four consecutive sky terraces.

Stainless steel columns of 20, 40, 60 and 80 cm diameter hang on steel bars off the ceiling. Some columns rise from the ground allowing them to enter the adjacent interior spaces. Under the hanging columns steel dishes with the same diameter will catch the water which drips down.

Some columns hang from the ceiling; others stand in space. The vertical, column-shaped gardens are stainless steel “cages” filled with substrate holding nutrients. Air and hydroponic plants (epiphytic ferns, Bromeliads, orchids, etc.) will cover their entire surface. The plants will be irrigated from inside the column and fed with air and moisture. Dripping water is solved by placing steel dishes of equal size under each column.

空中花园

空中花园是一系列矩形景观空间（长18米，宽8.4米），举架高达11.6米。每个空中花园将三个连续的楼层连接起来，拥有眺望城市风景和天际线的完美视野。从外面的街道上几乎看不到空中花园内部，它的风景只能从周围的室内空间来欣赏。

设计师打造了4个由线性元素构成的垂直花园，体量相当可观。观者可以近看空中花园中的植被，仔细欣赏每个细节。这4个空中花园分别位于11楼、23楼、33楼和45楼，每个花园内种植的植物都不同，所以可以从色彩、质地和香味上区分开来。

一楼的“热带花园”和冷却塔的绿化（都是绿墙的形式）以“柱式垂直花园”的形式复现于4个空中花园中。

不锈钢柱（直径分别为20厘米、40厘米、60厘米和80厘米）从天花板上的钢梁上悬垂下来。还有些钢柱从地面拔地而起。悬垂钢柱的下方设置了相同直径的钢盘，收集钢柱上滴下的水。

有些钢柱从天花垂下，有些则直立于地面，别具情趣。这种“柱式花园”的结构类似于不锈钢笼子，里面是含有养分的基质。气生植物和水培植物（附生蕨类植物、凤梨科植物、兰科植物等）将包裹住笼子表面。植物从笼子内部进行灌溉，供给所需的空气和湿度。滴水问题通过在悬垂钢柱下方放置同样大小的钢盘得以解决。

Sportplaza Mercator

墨卡托体育中心

Location:
Amsterdam, The Netherlands
Architect:
VenhoevenCS
Architecture+Urbanism
Landscape architect:
OKRA
Photographer:
Luuk Kramer, Copijn,
VenhoevenCS
Area:
7,100 sqm

项目地点：
荷兰，阿姆斯特丹
建筑设计：
维尔霍文建筑事务所
景观设计：
OKRA景观事务所
摄影师：
卢克·克莱默、Copijn景观事务所、维尔霍文建筑事务所
面积：
7,100平方米

Project description:

De Baarsjes in Amsterdam is a multicultural neighbourhood that is home to people from 129 different countries. The city district wanted to boost community life in the neighbourhood. The authorities therefore chose a building which combines swimming pools, a therapy pool, fitness, aerobics, a sauna and steam bath, a party centre, café and childcare alongside a fast food restaurant (jobs for the unemployed in the neighbourhood). Each individual element attracts different target groups, so the entire population will be able to use it in the end. Inside, everyone can see other activities, intriguing their interest and inspiring them to use other facilities as well. Because the building was constructed in a park, people living nearby it requested that it would be as green as possible; it was completely covered in vegetation.

Now, with its green façades and roof, Sport plaza Mercator marks the start and end of the Rembrandtpark. From a distance, it seems like an overgrown fortress flanking and protecting the entryway to the 19th-century city. Glimpsed through the glass façade, a modern spa-style complex glistens, complete with swimming pools, fitness space, and restaurant and party facilities. The entrance seems like a departure hall from which the various visitors can reach their destination.

The building was designed as a city – a society in miniature – inside a cave. The building is full of lines of sight and keyholes that offer perspectives on the various visitors, activities and cultures in the building. Sunlight penetrates deep into the building's interior through all sorts of openings in the roof. Low windows frame the view of the street and the sun terrace.

wemmen
&
Fitness
ClubAqua is o.a.:
AquaSpinning
uniek in A'dam!
Aquarobics
Aquajogging
banen zwemmen

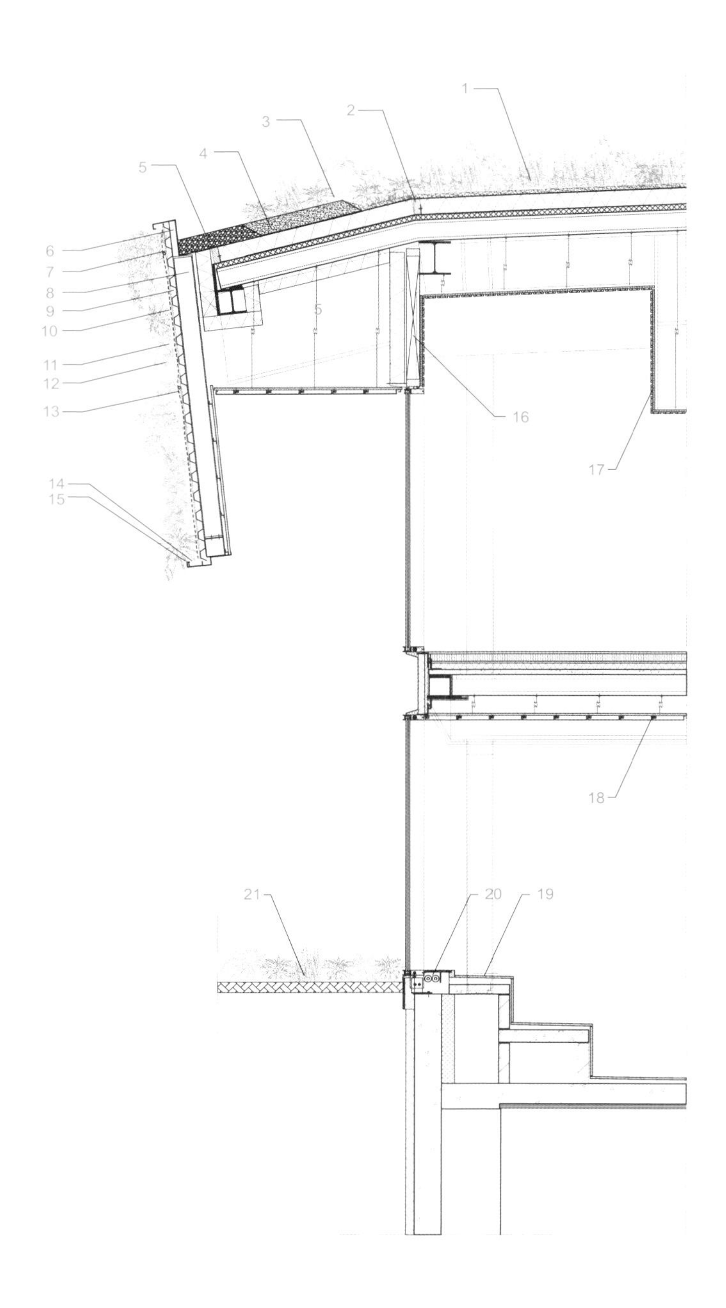

SECTIONAL DETAIL

1. Sowed sedum grasses
2. 40mm substrates
3. Permanent plants 6 per m^2
4. 150mm substrate
5. 150mm gravel
6. Tube 60x60x6mm
7. Water system, drip pipe
8. Columns green façade
9. Sheet-pile wall
10. Synthetic plate 10mm
11. 3 layers of textile & foil
12. Plants
13. Water system, drip pipe
14. Waste pipe
15. Steel frame
16. Wooden frame construction
17. Wooden lath
18. Expanded metal ceiling
19. Ivy plants
20. Convector
21. Granite floor

剖面大样图

1. 景天属植物
2. 生长基底（40毫米）
3. 永久性植物（每平方米6株）
4. 生长基底（150毫米）
5. 砾石（150毫米）
6. 管（60毫米×60毫米×6毫米）
7. 灌溉系统（滴管）
8. 植被包裹的柱子
9. 墙面
10. 合成板材（10毫米）
11. 织物和金属薄片（3层）
12. 植物
13. 灌溉系统（滴管）
14. 排水管
15. 钢质框架
16. 木质框架
17. 木板条
18. 延伸出来的金属天花板
19. 常春藤
20. 对流散热器
21. 花岗岩地面

ILLUSTRATION 效果图

阿姆斯特丹的杜巴罗斯区是一个多文化聚集区，居民来自129个不同的国家。这一地区的建设希望为居民提供更好的社区生活环境。墨卡托体育中心就是为这一目的而建，里面有游泳池、水疗池、健身房、有氧运动中心、桑拿和蒸汽浴房、派对中心、咖啡厅、儿童托管中心以及快餐店等（这意味着为这一地区创造出很多工作机会）。每项功能都面向不同的目标群体，所以，可以说全区居民都可以利用这里。体育中心内，每个人都可以看到其他的活动场景，说不定会激发起兴趣，忍不住参与其中。因为这栋建筑物建在一个公园里，所以住在附近的居民提出要求，要让这栋建筑物尽量绿化。设计师用植被将建筑表面彻底覆盖起来。

墨卡托体育中心的外立面和屋顶完全被绿植覆盖，完美地融入公园的绿色环境中。从远处看，这栋建筑就像一座自然生长的堡垒，保护着阿姆斯特丹这座始建于19世纪的老城。透过玻璃外立面向内窥视，现代水疗中心风格的场馆让人眼前一亮，游泳池、健身房、餐厅和派对场所一应俱全。入口门厅看起来像机场候机大厅一样，人们从这里可以直接到达他们想要活动的场所。

这栋建筑物设计得就像一座城市，一个掩映在绿色洞府中的微型社会。建筑物表面有很多线形和孔洞开口，从外面就能看到人们在室内参加各种活动的场景。屋顶上也有各式开窗，阳光能够深入建筑内部。透过立面上低矮的开窗，能看到街道和日光平台上的景象。

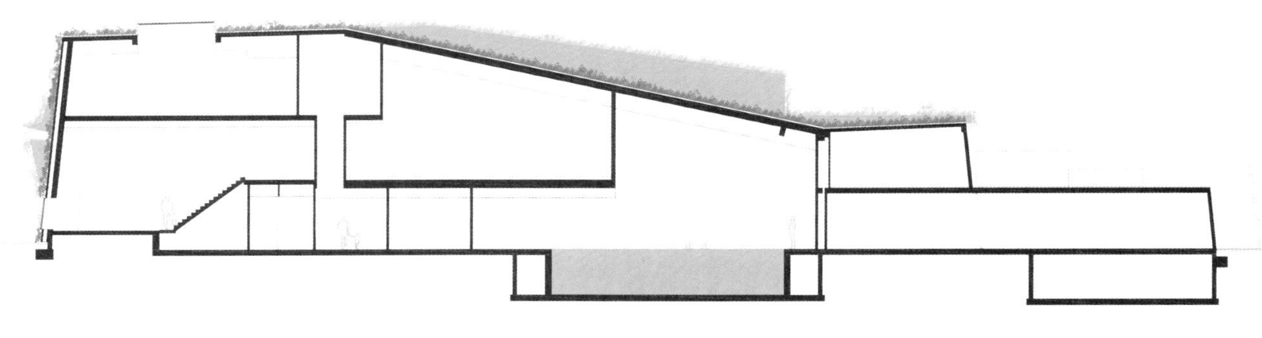

SECTION AA　剖面图AA

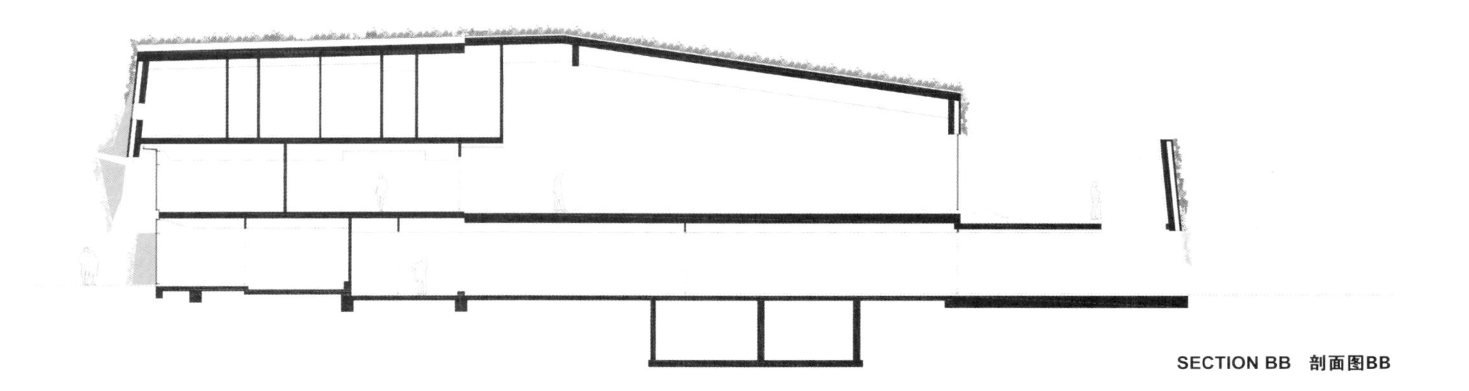

SECTION BB　剖面图BB

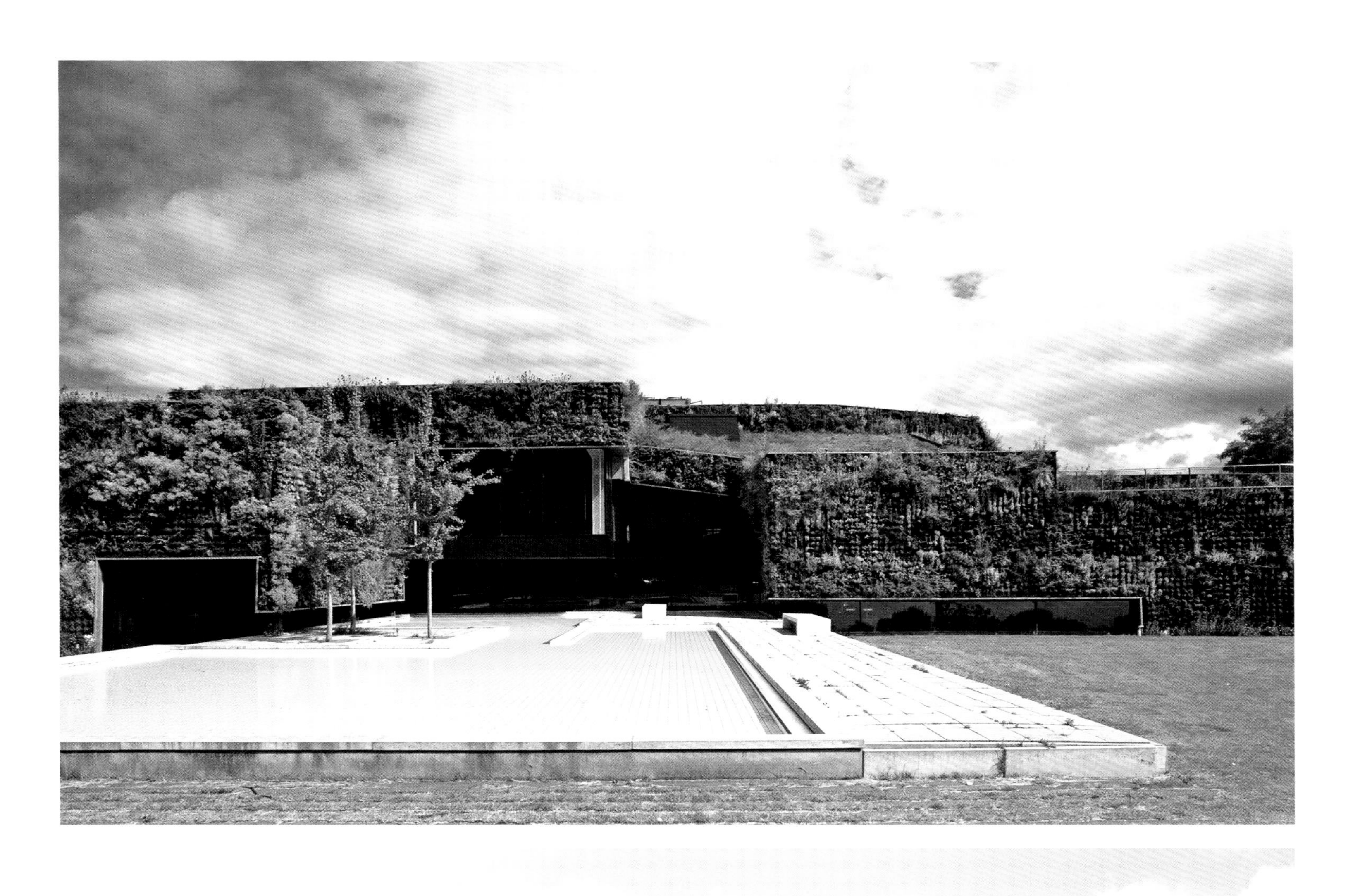

Sweet Tea, Beirut

贝鲁特甜茶餐厅

Completion date:
2011
Location:
Beirut, Lebanon
Architect:
Green Studios
Photographer:
Green Studios
Area:
110 sqm

竣工时间：
2011年
项目地点：
黎巴嫩，贝鲁特
景观设计：
绿色工作室
摄影师：
绿色工作室
面积：
110平方米

Project description:

Sweet Tea is the first one amongst restaurants in Beirut that includes a green wall installation; the restaurant's concept is an open sky terrace with four green walls.

Green Studios was appointed as the main consultant and contractor for the 110-square-metre green walls covering all four façades of the restaurant.

The scope of the project covers site evaluation, design development, detail drawings, plants selection and execution.

The interesting part of the design is the reverse promenade, where the visitor, who normally transits from an outdoor garden to an indoor space is actually experiencing the reverse process; as you walk within the souks, you exit the urban to this tea shop where the walls are green and the sky is wide open and your senses automatically enter into a different relaxed mood; the sound of water adds on to the Zen character of the space and at night, the whole space is dimly lit on purpose so the visitor could somehow disconnect from the surrounding urban jungle.

The main challenges for the green wall were that it had both an indoor part with low light, and an exposed outdoor one. There was also a tent that covered the area compromising the amount of light that the green wall gets during the day, which puts the whole installation under constant stress.

In order to seek fluidity between outdoor and indoor area, the designers paid great attention to the selection of plants which include a large amount of species. They are harmoniously combined together to create lively walls. The whole space resembles a vertical garden, which is a perfect space for visitors enjoying their nice time here. In addition to this, the designers apply advanced technology to maintain the plants.

甜茶餐厅是贝鲁特第一家采用室内绿墙的餐厅。甜茶餐厅的设计理念是四面绿墙围合而成的空中花园。

绿色工作室负责这家餐厅的绿墙设计，既是主要顾问，同时也是承包商。餐厅内四个立面的绿墙面积共计110平方米。

本案中设计师的工作范围包括：场地评估、设计开发、制图、选择植物和施工。

设计中最有趣的部分就是空间体验的反转。一般来说，顾客是经过外面的花园进入室内空间，而这里则正好相反。你走过露天市场来到这家小餐厅门前，进入餐厅，好像从喧闹的都市环境一下迈进一个绿色的世界。这里，四周的墙上长满植物，头顶的天空开阔蔚蓝，你的所有感官不知不觉进入了另一种放松的状态。哗哗的流水声为空间增添了一丝禅意。夜间，整个空间笼罩在朦胧的灯光下，有意让顾客从感觉上远离外部喧嚣的城市环境。

绿墙设计面临的最大挑战是：既有室内部分（光照不足），也有室外部分（暴露在阳光下）。上方有遮棚可以控制绿墙上的植物白天得到的光照。

为了营造室内和室外空间之间的流畅感，设计师在植物的选择上颇费心思，选择了大量不同的品种。这些品种通过巧妙的搭配，共同营造出欣欣向荣的绿墙。整个空间仿佛一座“垂直花园”，顾客在这里可以轻松度过一段悠闲的时光。此外，设计师还采用了先进的技术，保证了植物未来的维护。

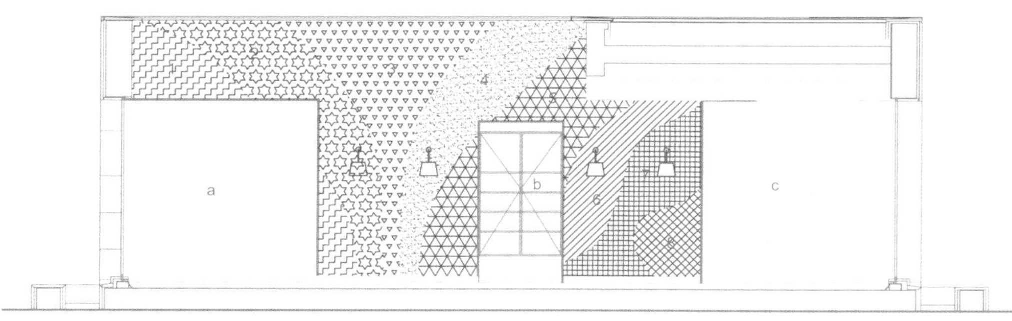

FRONT WALL

a. Fixed glass window to the outside
b. Food exposition cold case
c. VIP tearoom
1. 2m^2
2. 3.6m^2
3. 3.8m^2
4. 3.3m^2
5. 2.6m^2
6. 1.7m^2
7. 2m^2
8. 1m^2

正面墙面立面图

a.玻璃窗固定在外部
b.食品展示冷藏柜
c.贵宾茶室
1. 2平方米
2. 3.6平方米
3. 3.8平方米
4. 3.3平方米
5. 2.6平方米
6. 1.7平方米
7. 2平方米
8. 1平方米

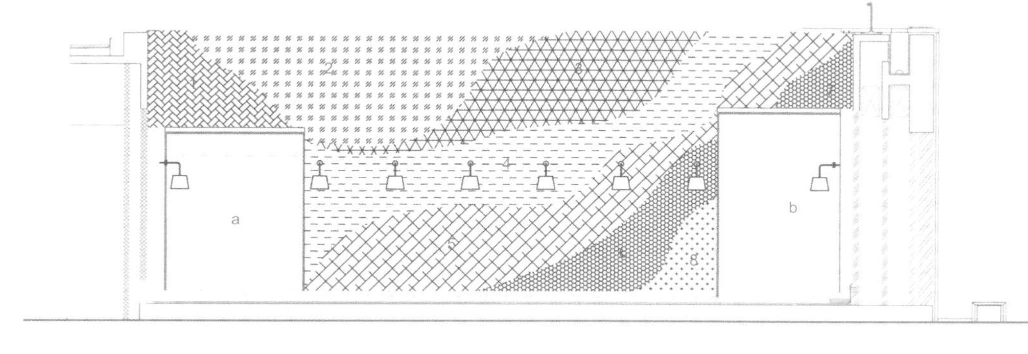

TERRACE WALL

a. Terrace access
b. Fixed glass window to the outside
1. 1.9m^2
2. 5.6m^2
3. 4.2m^2
4. 6.5m^2
5. 6.4m^2
6. 2.3m^2
7. 0.7m^2
8. 0.9m^2

平台墙面立面图

a. 平台入口
b. 玻璃窗固定在外部
1. 1.9平方米
2. 5.6平方米
3. 4.2平方米
4. 6.5平方米
5. 6.4平方米
6. 2.3平方米
7. 0.7平方米
8. 0.9平方米

Vertical Garden at Palace of Congresses

国会宫垂直花园

Completion date:
November 2013
Location:
Vitoria-Gasteiz, Spain
Landscape architect:
Urbanarbolismo &UnusualGreen
Photographer:
Hugo y Jordi
Area:
1,492 sqm

竣工时间：
2013年11月
项目地点：
西班牙，维多利亚
景观设计：
"城市景观"事务所、"非常绿化"事务所
摄影师：
雨果+霍尔迪
面积：
1,492平方米

Project description:

The vertical garden system "f + p preplant" is made with a non-woven fabric that distributes water over the façade. Planted rockwool panels are placed on the non-woven fabric. This process warranties no stress for the plant and cold resistance.

The city of Vitoria-Gasteiz is recognised internationally for integrating biodiversity and green spaces in the urban architecture. Within the Inside Green Ring Project Urbanarbolismo and UnusualGreen studios in collaboration with Urbaser and Zikotz have made this green façade to bring the surrounding ecosystems to the city centre.

The vertical garden has a total area of 1,492 square metres, of which 1,000 square metres are a hydroponic vertical garden system "f + p" and 492 square metres are made of climbing plants covering the windows. More than 33,000 native plants of different varieties from Alava and the Basque Country have been used.

The main reason to make the project was to improve energy consumption of the Palace of Congresses. The vertical garden "f + p preplant" system adds a thermal resistance of 2,644 m^2K/W. This represents a 270% increase on the existing façade insulation, resulting in energy savings.

The design performs the ecosystems that exist in the surrounding of the city of Vitoria-Gasteiz. From left to right the façade shows the wetland vegetation of Salburua, the agricultural field plots of Alava, the ecosystems from the loamy hills and beech forests of the mountains of Vitoria.

At the base of the façade a Corten steel back-

EUROPA

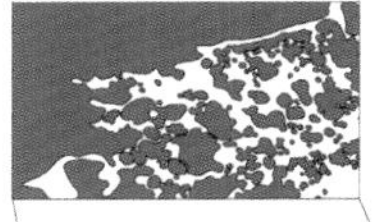

Structural characteristics of clusters of vegetation in wetlands Salburua
西班牙Salbar ú a湿地植被群的特征结构

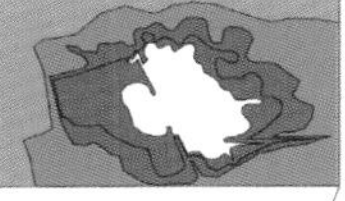

Transition of vegetation typologies from the centre to the edge of the wetland
从湿地中心到边缘的植被种类过渡

Vitoria farm field structure
维多利亚地区的农田结构

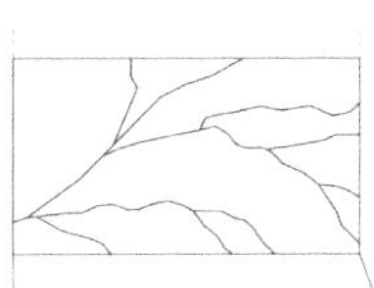

Hydraulic structure of the mountains of Vitoria
维多利亚地区山上的水结构

Vertical garden, 101sqm
The use of an inert substrate seen allows mosses and plants colonise the surface creating a wallpaper background for the rest of the plantation.
Low maintenance
Water consumption: 3litres/sqm/day

垂直花园（101平方米）
外墙表面培育层的使用让苔藓等植被得以生长，并在植物和植物之间形成了一道天然的"壁纸"
低维护
耗水量：3升/平方米/天

Vertical wetland, 208sqm
A selection of species associated with Vitoria wetlands planted one to one forming a structured design wetland.
Normal maintenance
Water consumption: 5litres/sqm/day

垂直湿地（208平方米）
维多利亚地区植被种类的选择要适应湿地的结构特征
正常维护
耗水量：5升/平方米/天

Vertical orchard, 5sqm
A small part of the garden reserved for an orchard: tomatoes, peppers, lettuce, cabbage, cauliflower... as part of the whole garden.
High maintenance
Water consumption: 4litres/sqm/day

垂直菜园（5平方米）
在花园中预留出来一部分空间，作为一个小型菜园，种植土豆、花椒、生菜、卷心菜等
高维护
耗水量：4升/平方米/天

Vertical patchwork, 482sqm
The vertical garden planting is done with sedum and turf grasses that reproduce the design of Vitoria agricultural field structure. No need of pruning.
Minimum maintenance
Water consumption: 1.4litres/sqm/day

垂直补缀（482平方米）
垂直花园的种植是和再造维多利亚地区农田结构的景天草草皮共同实现的，无需修枝
最低维护
耗水量：1.4升/平方米/天

Climbing plants, 205sqm
Selection of deciduous climbing plants protects from hollow summer sun and allows the entry of light in winter. Fastened on a structure of stainless steel tensioned cables.
Normal maintenance
Water consumption: 2litres/sqm/day

藤蔓植物（205平方米）
所选择的藤蔓植物不仅可以有效地阻挡夏日里强烈的阳光，还能在冬日里让阳光照射进来。这些藤蔓植物是由许多不锈钢弹簧结构作为支撑的
正常维护
耗水量：2升/平方米/天

Vertical forest, 269sqm
Selection related to the flora of the mountains of Vitoria that reproduces a structure of several species.
Normal maintenance
Water consumption: 3litres/sqm/day

垂直森林（269平方米）
垂直森林的种类要根据维多利亚山区植物结构进行选择，垂直森林再造了一种河流结构
正常维护
耗水量：3升/平方米/天

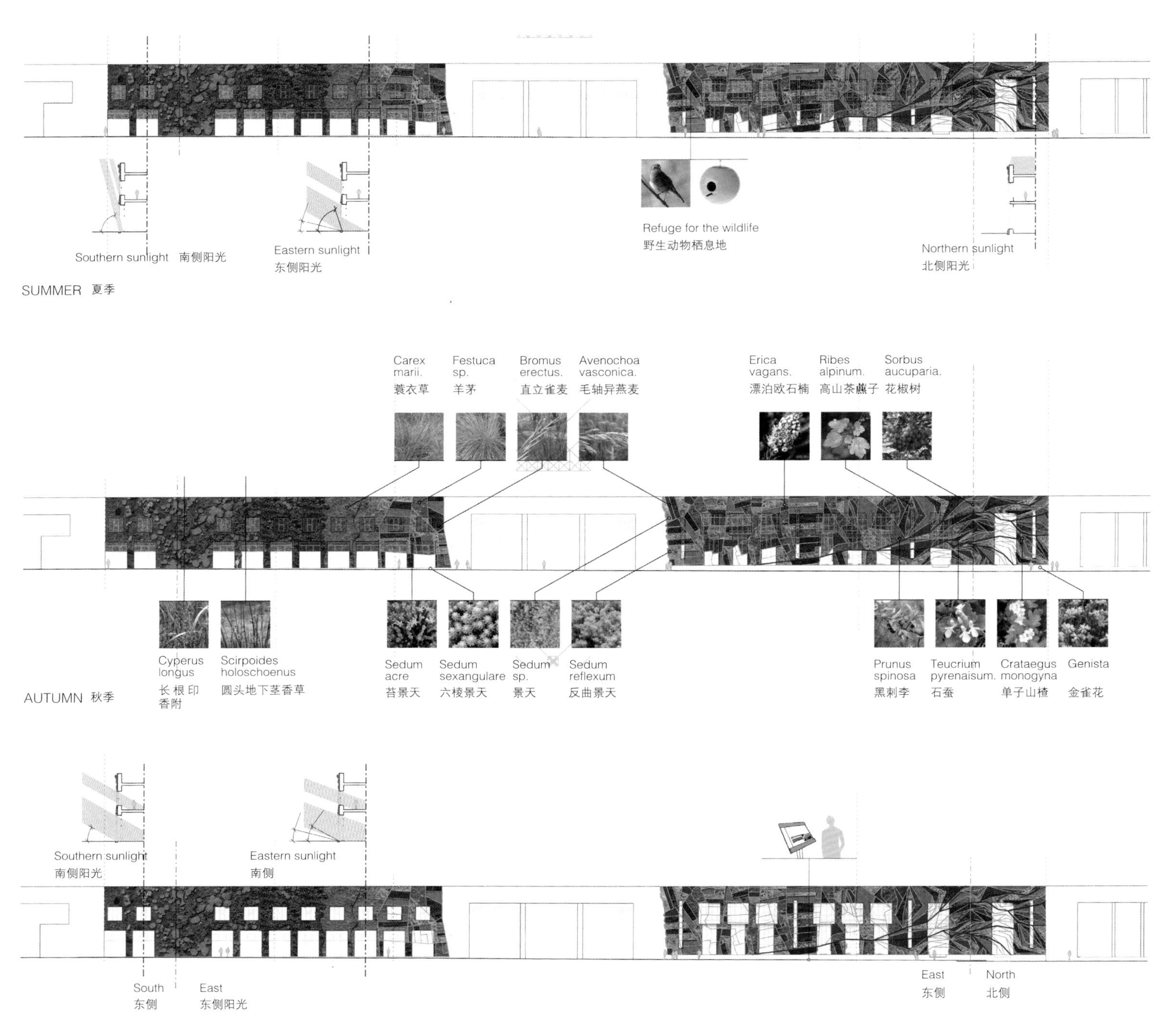

lit socket have been placed, which contains an explanation of the ecosystems and plant species of the vertical garden. 97% of the species used in the garden are native or endemic from Alava. This is one of the first vertical garden reproducing native ecosystems of the area where it is located.

The use of native plants in the design has been one of the main challenges of the project. Many of the plants from Vitoria are adapted to periods of drought and are struggling to survive in moist environments such as conventional vertical gardening systems. For this project, the "f+p" system was redesigned, optimising substrate saturation so that these plants can perfectly grow.

The hydroponic system used to maintain this garden gives the optimal substrate nutrient conditions, pH, conductivity and humidity for this type of vegetation. The whole system is monitored by remote control in order to save water, energy and supervise the development of plants.

Starting the tour of the vertical garden from left to right we first find the Vertical Wetlands. The façade reproduces the formal pattern of Salburua wetlands (a wetland area near Vitoria- Gasteiz). The species used are typical of wetlands in the area: Scirpoides holoschoenus, Cyperus longus, Carex mairii, Juncus maritimus, Juncus acutus, Cirsium monspessulanun, Tetragonolobus maritimus, Lysimachia ephemerum...

本案中的垂直花园采用“F+P”预种植系统，以无纺布为基础材料，将水分均匀分布到整个外立面上。无纺布上安装石棉板，上面种植植物。这种结构能够确保植物不会承受压力，也有利于植物御寒。

西班牙维多利亚市以生物多样性和城市环境绿化在国际上享有盛名。本案是维多利亚“绿环”工程的一部分，由“城市景观”事务所和“非常绿化”事务所两家公司联手Urbaser废物处理公司和Zikotz建筑工程公司共同完成。这面绿墙让维多利亚附近的生态系统回归城市中心。

绿墙面积共计1,492平方米，其中包括1,000平方米的水培植物（也就是“F+P”预种植系统）和492平方米的攀爬植物（覆盖住窗户）。绿墙上采用的植物种类繁多，包括33,000多种本地植物，原产于阿拉瓦省和巴斯克自治区。

启动本案的主要原因是为了改善国会宫的建筑能耗。“F+P”预种植系统相当于在建筑外立面上加上热阻为2,644 m^2K/W的隔热层。这表示原建筑外立面的隔热作用提升了270%，节约了大量能源。

绿墙的设计模拟维多利亚市周围的自然生态系统。从左到右，外立面上依次呈现出萨尔布鲁阿湿地植被、阿拉瓦省农田、肥沃山脉上的生态系统以及维多利亚山上的山毛榉树林。

绿墙下方有一块耐候钢材质的基座，上面的文字介绍了绿墙生态系统和植被种类。绿墙中用到的97%的植物都是阿拉瓦省的本地植物。本案是当地首批本地生态系统垂直花园之一。

本地植物的使用是本案中的一大难题。维多利亚植物需要适应干旱期，还必须能在潮湿的环境下生存（一般的绿墙系统都是比较潮湿的生长条件）。设计师专门为本案而改造了“F+P”预种植系统，让培养基具有更高的吸水饱和度，以便植物能更好地生长。

In the centre of the façade aluminium profiles reproduce the structure of the field plots of Alava. In this space the landscape architects have allocated an area for a Vertical Orchard. In winter cress and lettuce were planted.

In the northern side of the façade the structure of profiles becomes more organic. This is the Vertical Forest. This is the place that plays the ecosystem of Vitoria hills and mountains. Some of the species used are: Ribes alpinum, Sorbus aria, Crataegus monogyna, Prunus spinosa, Erica vagans, Genista occidentalis, Globularia nudicaulis, Teucrium pyrenaicum, Bromus erectus, Carex humilis... Even two species of tree have been planted: Beech "Fagus sylvatica" and yew "Taxus baccata".

Large windows will be covered by deciduous vines that allow passage of sunlight in winter and protect the building from the summer heat, creating a biologically eco-efficient building. The river-profile that structures vertical forest design is illuminated at night by low-power LEDs. This creates a unique atmosphere for a walk in the centre of Vitoria-Gasteiz.

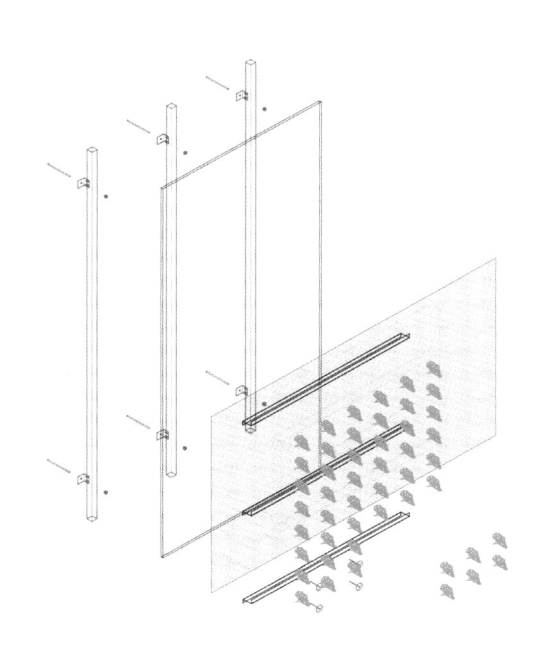

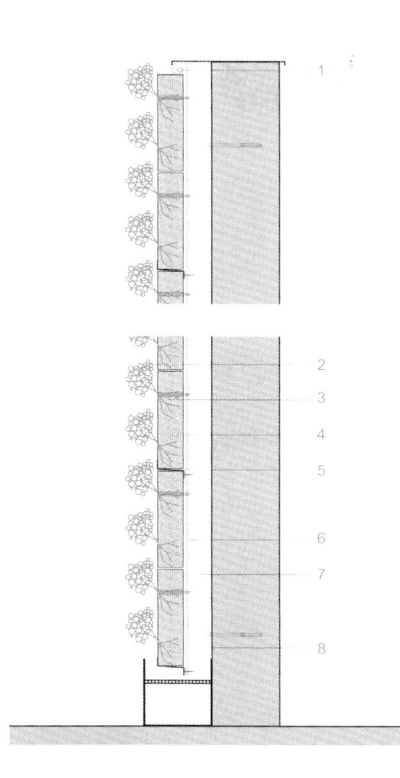

VERTICAL GARDEN DETAIL

1. Porous irrigation Ug-R16
2. Especially selected plants growth in the nursery for two months
3. Non-woven fabric Ug-M500
4. Pre-plant Panel rockwool Ug-L50-100 600x300mm
5. Aluminium plattern 2mm Ug-Pl75
6. Panel Ug-P10
7. Aluminium frame 40x40.3mm
8. Anchor bracket stainless steel

垂直花园细部详图

1. 渗透性灌溉Ug-R16
2. 精选植物（已经在花圃中培育两个月）
3. 无纺布Ug-M500
4. 石棉Ug-L50-100（600毫米×300毫米）
5. 铝板Ug-Pl75（2毫米）
6. 板材Ug-P10
7. 铝框（40毫米×40.3毫米）
8. 不锈钢锚架

The Corten steel base not only provides educational information on the garden, but also becomes a space for relaxation and conversation, a bench in a Vertical Park. Quickly nature breaks through the vertical garden. Thanks to native plants, it has been colonised by all kinds of insects and animals. Birds have quickly appeared. Here you have a picture of a magpie foraging.

本案绿墙采用的水培系统能够为植物提供最佳的营养条件、pH值、导电性和湿度。整个系统能够远程控制，实现节水、节能，并且能够监测植物的生长情况。

按照从左到右的顺序“游览”这座“垂直花园”，首先你会看到“垂直湿地”。绿墙上再现了萨尔布鲁阿湿地的生态景观（萨尔布鲁阿是离维多利亚不远的湿地）。绿墙所用的植物都是萨尔布鲁阿湿地的常见植物，包括蔺木贼、莎草、蓑衣草、矮灯芯草、尖灯芯草、蓟草、翅荚豌豆、金钱草（珍珠菜）等。

绿墙中间是铝制框架，模拟的是阿拉瓦省的农田景观。设计师在这个部分打造了一座“垂直果园”，在冬季时栽种了山芥和莴苣。

绿墙北侧的植被呈现出有机形态。这里是“垂直森林”。这里的植被模拟维多利亚山的自然生态系统。采用的植物包括：高山柳叶菜、白面子树、单子山楂树、黑刺李、漂泊欧石楠、欧美金雀花、球花、石蚕、直立雀麦、低矮苔草等。这里甚至还栽种了两种树木——欧洲山毛榉和欧洲紫衫。

外立面上巨大的开窗爬满落叶藤蔓植物。冬季，阳光可以射入；夏季起到遮光的作用，完善了建筑物的生物效能。“垂直森林”中有模拟河流的纵横结构，夜晚采用低能耗的LED灯进行照明。夜间在维多利亚市中心散步，你便会感受到这里独特的氛围。

耐候钢告示板不仅起到信息宣传的作用，而且化身为长椅，成为这座“垂直花园”里的一个休闲、谈话的所在。大自然很快便回归城市。由于绿墙上都是本地植物，所以各种昆虫和动物很快在这里安家。鸟儿也很快出现了，绿墙上呈现出如画般的喜鹊觅食景象。

Canan Residence

卡南住宅

Location:
Istanbul, Turkey
Landscape Project:
DS Architecture – Landscape
Landscape architect:
Deniz Aslan
Architectural Project:
MAM - Adnan Kazmaoğlu
Client:
Canan Yapı Group
Photographer:
Gürkan Akay, Deniz Aslan
Area:
10,000 sqm

项目地点：
土耳其，伊斯坦布尔
景观公司：
DS建筑事务所
景观建筑师：
德尼兹・阿斯兰
建筑公司：
MAM–艾德楠・卡兹莫格鲁
客户：
卡南雅蒲置业集团
摄影师：
戈尔坎・阿凯、丹尼斯・阿斯兰
面积：
10,000平方米

Project description:

Among the three white towers, landscape design intends to create an unadorned garden which offers various kinds of experiences, and yet it considers this simple garden that shelters different activities within its frame, as an extension of the architectural scenario. In Canan Residence, the garden adapted itself to the absolute dialect of the architectural style. In other words, landscape correlated with architecture and thus an architectonic interplay occurred as a result. In terms of used materials, while the white concrete was dominant on the construction of the structural elements, white andesite which was adopted by the architectural project, was chosen for the walking paths. The garden was designed simply as a means that ascribe different opportunities to the blank spaces among vertical connections. In this regard, the design challenge as an intention to create a slight but well effective garden experience turned into an educative process for the design team.

The most dominant activity of the terrace garden and the entire project zone, is the sunken water garden that shelters a social facility which brings out a recreational centre. The design was built up in a linear sequence; a swimming pool, a kid's pool, an artifact of mist, and a simple and statuesque water garden. Except for the remaining space that is used for sunbathing and social activities, the garden was designed almost without any plantation. However, this unique characteristic of the landscape design was not considered odd at any point.

Parallel to the water garden, the upper elevation embodies a row of pergolas, which among its load-bearing axes provides a ground for the pocket gardens. On the other hand, this well-settled structure balances the overall scale of the project itself. Linear seating elements which are used in the entire area, facilitated the organisation of sports area's viewing platforms and also provided

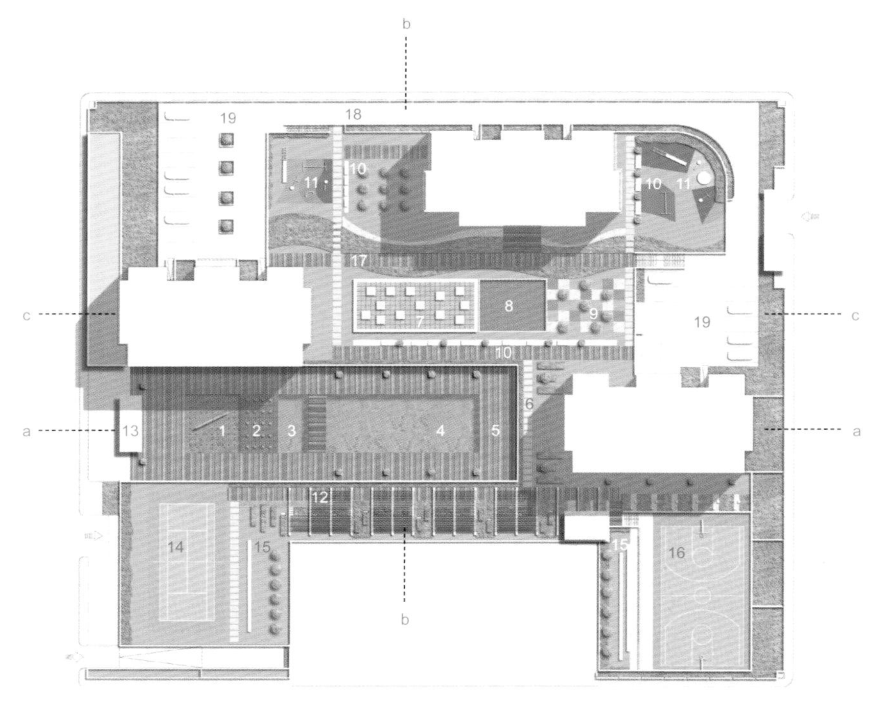

PLAN

1. Reflection pool
2. Water labyrinth
3. Children's pool
4. Swimming pool
5. Sunbathing wooden deck
6. Stepping stone path
7. Skylight
8. Reflection pool
9. Resting area
10. Sitting furniture
11. Children's playground
12. Canopy & wooden deck
13. Bar canopy
14. Tennis court
15. Viewing platform
16. Basketball court
17. Wooden walkway
18. Road
19. Parking area

平面图

1. 倒影池
2. 水景迷宫
3. 儿童泳池
4. 泳池
5. 木质日光浴平台
6. 石头小径
7. 天窗
8. 倒影池
9. 休息区
10. 坐具
11. 儿童游乐区
12. 遮棚与木质平台
13. 酒吧遮棚
14. 网球场
15. 观景平台
16. 篮球场
17. 木质走道
18. 道路
19. 停车区

a transition plane between various functions. Consisting of simple orthogonal geometries, the linear garden above the social (recreational) facility, gave way to the emanation of a cubic plasticity. The glass cubes that engender the most striking pattern of the planometric design, act as the brutal reflection of the load-bearing system and hence, they enable the enclosed pool to get daylight, up above from the ceiling. Along with their statuesque stance and functional presence, these design elements brought to garden a new dimension. In addition, the concept of soft lighting especially for night landscape helped the creation of a meaningful unity.

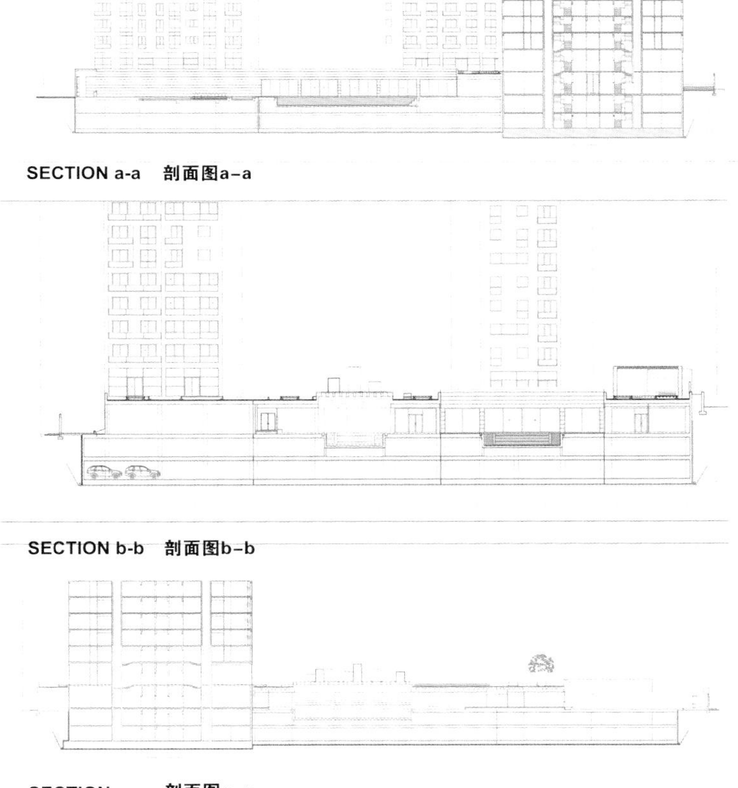

SECTION a-a 剖面图a–a

SECTION b-b 剖面图b–b

SECTION c-c 剖面图c–c

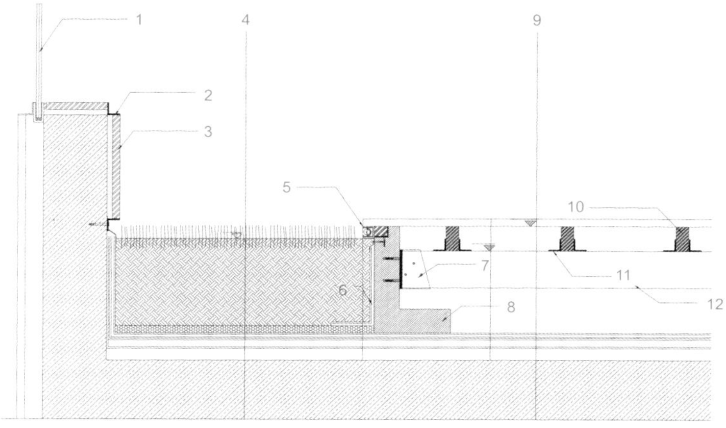

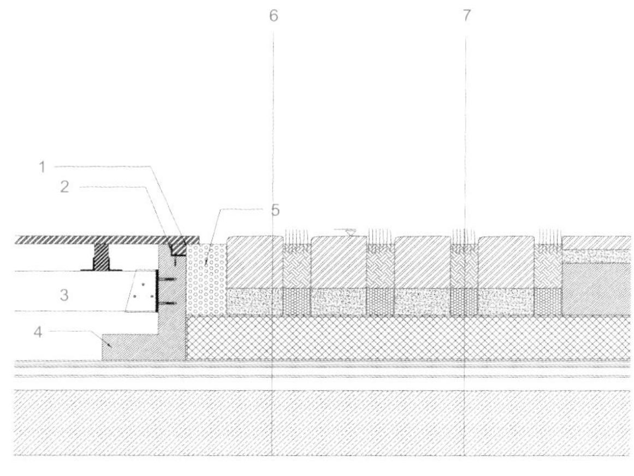

WOOD DECK DETAIL

1. Glass balustrade
2. Metal profile 5/5
3. Natural stone
4. Planting medium 35cm
   Pumice stone 3cm
   Geotextile
   Drainage membrane 1cm
   Root barrier
   Geotextile
   Xps
   Water isolation
   Levelling concrete 4-6cm
   Concrete slab
5. Linear LED lighting
6. Geotextile drainage membrane 500gr/m^2
7. Metal anchoring
8. Concrete footing
9. Wooden deck 3cm
   Wooden timber 5/10
   Wooden framing 10/15
   Geotextile
   Drainage membrane 1cm
   Root barrier
   Geotextile
   Xps 3cm (ext. polystyrene)
   Water isolation
   Levelling concrete 4-6cm
   Concrete slab
10. Wooden timber
11. Metal L profile 5/5
12. Wooden beam 10/15

木质平台细部图

1. 玻璃扶栏
2. 金属材料（5/5）
3. 天然石材
4. 种植介质（35厘米）
   浮石（3厘米）
   土工织物
   排水膜（1厘米）
   根系屏障
   土工织物
   挤塑板
   防水层
   混凝土找平层（4～6厘米）
   混凝土板材
5. 线性LED照明
6. 土工织物排水膜（500gr/m^2）
7. 金属固定装置
8. 混凝土底脚
9. 木质平台（3厘米）
   木材（5/10）
   木框（10/15）
   土工织物
   排水膜（1厘米）
   挤塑板（3厘米，聚苯乙烯材质）
   防水层
   混凝土找平层（4～6厘米）
   混凝土板材
10. 木材
11. “L”形金属材料（5/5）
12. 木梁（10/15）

GRANITE CUBE PAVEMENT DETAIL

1. Wooden timber 4x5
2. Metal L profile
3. Wooden beam 10/15
4. Concrete footing
5. White gravel
6. Granite cube stone 20x20x20cm
   Mortar 10cm
   Rubble
   Geotextile
   Drainage membrane 1cm
   Root barrier
   Geotexile
   Xps 3cm (ext. polystyrene)
   Water isolation
   Levelling concrete 4-6cm
   Concrete slab
7. Planting medium 35cm
   Pumice stone 3cm
   Rubble
   Geotexile
   Drainage membrane 1cm
   Root barrier
   Geotextile
   Xps 3cm (ext. polystyrene)
   Water isolation
   Levelling concrete 4-6cm
   Concrete slab

花岗岩铺装细部图

1. 木材（4x5）
2. “L”形金属材料
3. 木梁（10/15）
4. 混凝土底脚
5. 白色砂砾
6. 花岗岩石块（20x20x20厘米）
   灰泥（10厘米）
   碎石
   土工织物
   排水膜（1厘米）
   根系屏障
   土工织物
   挤塑板（3厘米，聚苯乙烯材质）
   防水层
   混凝土找平层（4～6厘米）
   混凝土板材
7. 种植介质（35厘米）
   浮石（3厘米）
   碎石
   土工织物
   排水膜（1厘米）
   根系屏障
   土工织物
   挤塑板（3厘米，聚苯乙烯材质）
   防水层
   混凝土找平层（4～6厘米）
   混凝土板材

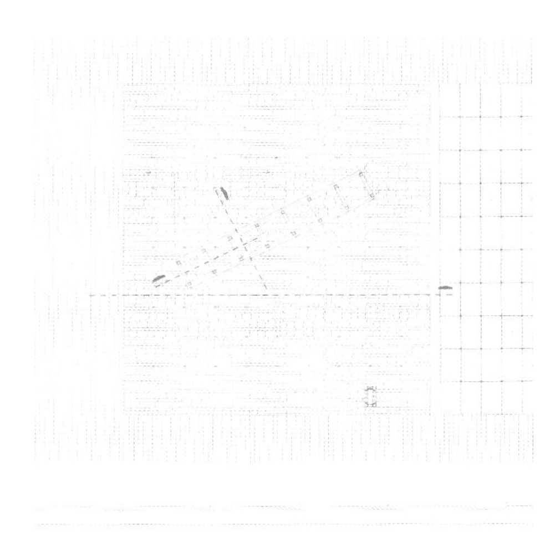

WATER FEATURE PLAN 水景平面图

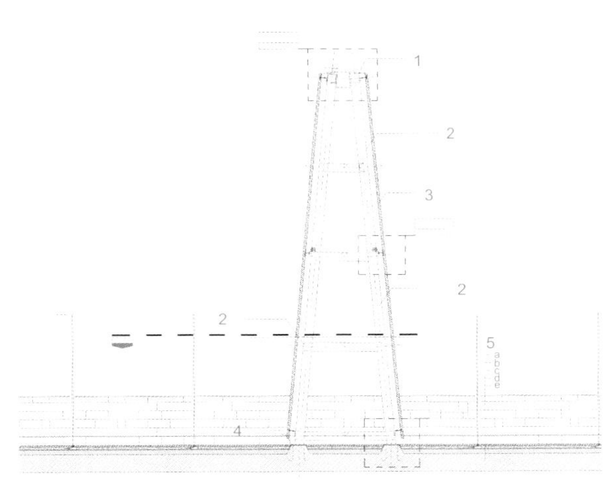

WATER FEATURE ELEVATION

1. Stainless steel water channel
2. 2XNPU 10 stainless steel profile
3. Drain Φ200mm
4. Natural stone 3cm
5. a. Diabase 2cm
   b. Mortar 3cm
   c. Water isolation
   d. Levelling concrete 4-6cm
   e. Concrete slab

水景立面图

1. 不锈钢水槽
2. 不锈钢材料（2XNPU 10）
3. 排水管（直径200毫米）
4. 天然石材（3厘米）
5. a.辉绿岩（2厘米）
   b.灰泥（3厘米）
   c.防水层
   d.混凝土找平层（4厘米～6厘米）
   e.混凝土板材

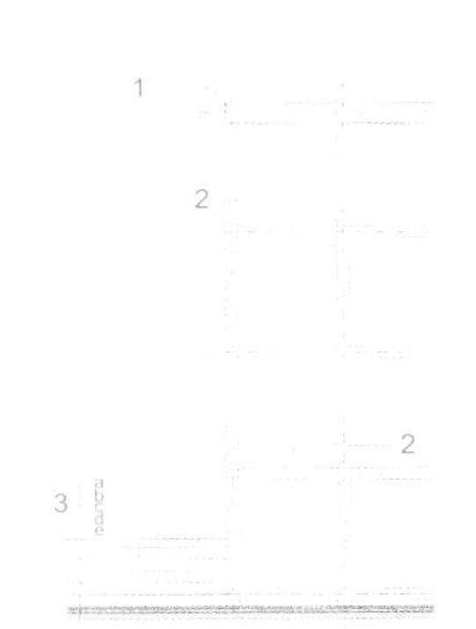

WATER FEATURE SECTION 2

1. Stainless steel water channel
2. 2XNPU 10 stainless steel profile
3. a. Diabase 2cm
   b. Mortar 3cm
   c. Water isolation
   d. Levelling concrete 4-6cm
   e. Concrete slab

水景剖面图-2

1. 不锈钢水槽
2. 不锈钢材料（2XNPU 10）
3. a.辉绿岩（2厘米）
   b.灰泥（3厘米）
   c.防水层
   d.混凝土找平层（4厘米～6厘米）
   e.混凝土板材

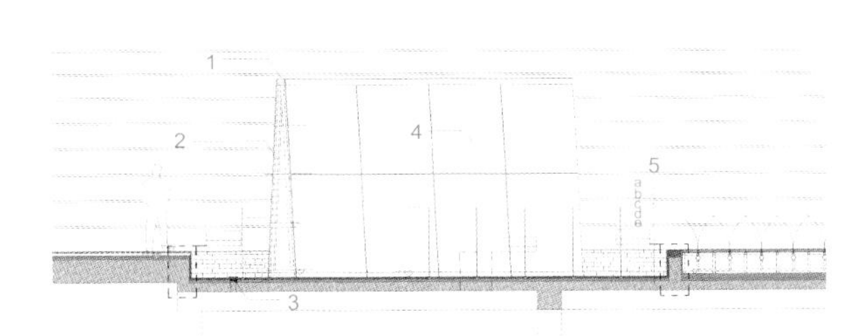

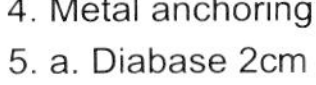

WATER FEATURE SECTION 1

1. Stainless steel water channel 10mm
2. 2XNPU 10 stainless steel profile
3. Natural stone 3cm
4. Metal anchoring plate 10mm
5. a. Diabase 2cm
   b. Mortar 3cm
   c. Water isolation
   d. Levelling concrete 4-6cm
   e. Concrete slab

水景剖面图-1

1. 不锈钢水槽（10毫米）
2. 不锈钢材料（2XNPU 10）
3. 天然石材（3厘米）
4. 金属固定板（10毫米）
5. a.辉绿岩（2厘米）
   b.灰泥（3厘米）
   c.防水层
   d.混凝土找平层（4厘米～6厘米）
   e.混凝土板材

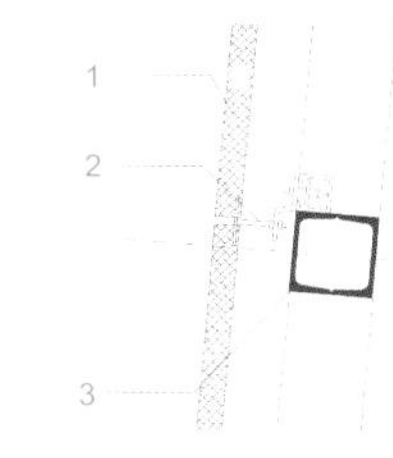

ANCHORING DETAIL

1. Natural stone
2. Stainless steel anchor
3. 2XNPU 10 stainless steel profile

固定装置细部图

1. 天然石材
2. 不锈钢固定装置
3. 不锈钢材料（2XNPU 10）

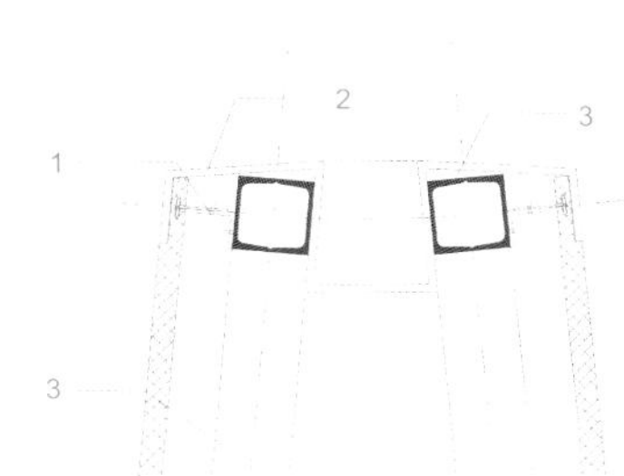

WATER CHANNEL DETAIL

1. Stainless steel anchoring
2. Stainless water channel
3. 2XNPU 10 stainless steel profile

水槽细部图

1. 不锈钢固定装置
2. 不锈钢水槽
3. 不锈钢材料（2XNPU 10）

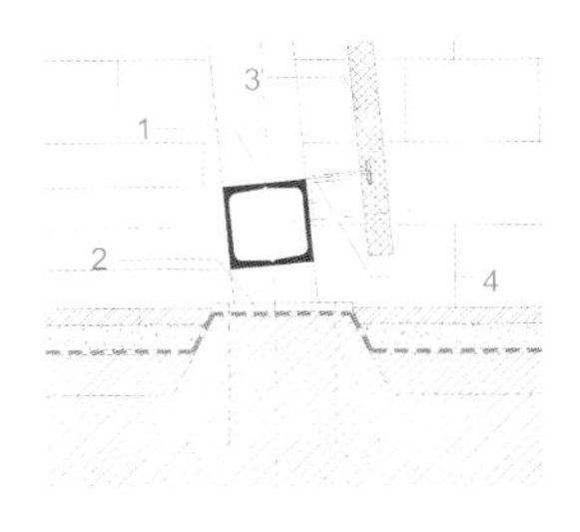

FOUNDATION DETAIL

1. 2XNPU 10 stainless steel profile
2. Stainless steel anchoring plate 10mm
3. Natural stone 3cm
4. Stainless steel anchoring

喷泉细部图

1. 不锈钢材料（2XNPU 10）
2. 不锈钢固定板（10毫米）
3. 天然石材（3厘米）
4. 不锈钢固定装置

As usual, the design that came into being on a terrace garden, pushed the team to prevail special precautions and formulas in order to protect plants; especially trees from the rough conditions of the project like, the strong wind, minimum height of soil, etc. Especially, the success of the intention of preserving the irrigation water with a specific drainage system, helped the maintenance of the landscape with a minimum amount of irrigation throughout the hot summer days, and yet the residents put emphasis on the importance of the evaporating water for the sustainability of plants life cycle.

卡南住宅的三栋白色大楼环绕着中央的台地花园。设计师希望打造一座朴素的花园，带给住户多样化的体验。花园虽简单，但是里面却可以进行丰富多彩的活动。花园作为建筑物的延伸，要与建筑融为一体。花园的设计采用与这三栋住宅楼相同的建筑语言。换句话说，花园与建筑是相辅相成的关系，二者相互影响，相互作用。在材料的使用方面，花园的结构框架大量采用白色混凝土，而小径的铺装则采用住宅建筑所用的白色安山岩。花园的设计目标非常简单，就是要在住宅建筑垂直结构的空隙中创造更多的可能性。从这个角度上说，设计师面临的挑战就是如何设计一个既简单又功能齐全、体验多样的花园，对设计团队来说，这也是一个学习的过程。

这个台地花园以及整个住区内最重要的活动场地就是下沉的水景花园。这是一个娱乐中心。空间布局采用线性设计，包括泳池、儿童泳池、人工“雾境”以及一个造型简单的水景花园。水景花园内除了进行日光浴和各种社交活动的空间之外，其他地方几乎没有绿化。不过，这一独特的景观设计却丝毫不显得古怪。

水景花园旁边有一排棚架与之平行，棚架的标高更高些，下方有几根承重轴，支撑着上方的一系列迷你花园。这种巧妙的结构消解了住宅楼巨大的体量。棚架下方是社交（娱乐）设施。水景花园的另一边设置线性座椅，让体育运动区的几个观景台的布局更加清晰，同时也是各种功能之间的一个过渡。这边是由直角几何结构构成的线性花园，体现出立体派的美感。玻璃立方体在花园的平面布局中凸显出来。立方体本身是承重结构，下面是封闭的泳池，阳光能够透过玻璃照射进去。这些雕塑般的立方体为花园增添了别具一格的设计元素。此外，“软照明”的理念（尤其是夜间景观照明）让整个花园笼罩在统一的氛围下。

由于景观设计在台地上进行，所以设计师照例采取了一些预防措施，确保植物正常生长，尤其是树木，因为场地的生长条件很恶劣（大风、土壤层很薄，等等）。设计师还特别注意灌溉用水的节约，专门设计了排水系统，确保在炎热的夏季只需最少的灌溉水量就能满足景观的维护需求，而居民则强调水池中水气蒸发有利于植物的可持续生长。

Giant Interactive Campus

巨人网络集团总部

Completion date:
2011
Location:
Shanghai, China
Architect:
Morphosis Architects
Landscape architect:
SWA Group
Photographer:
Tom Fox
Area:
182,108 sqm

竣工时间：
2011年
项目地点：
中国，上海
建筑设计：
"形态"建筑事务所
景观设计：
SWA景观设计集团
摄影师：
汤姆·福克斯
面积：
182,108平方米

Project description:

The 18ha campus and green roof for Giant Interactive Group in Shanghai, China was conceived as an ecological park and living laboratory. Structured around a plan of natural systems and open space, both the landscape and architecture programmes seamlessly integrate. Half of the campus contains corporate and office uses, while the other half is focused on lifestyle, and includes a hotel, clubhouse, along with dining and recreational spaces.

SWA sought to use the landscape as an organising framework for the master plan with a variety of water experiences focused on ecological sustainability. The site was formerly a tree nursery for mature Camphors (Cinnamomum camphora) and Sweet Olives (Osmanthus fragrans) surrounded by agricultural canals. Many of the existing trees were salvaged and relocated across the campus.

A city road (Zhongkai Road) bisects the land, splitting it into two sections. Planning of the site employed water as a means to connect and organise various programme elements. An intricate hydrological system consisting of existing irrigation canals, new water retention basins, islands, and seasonal wetlands, created a diverse habitat for wildlife as well as scenic points. Water and wetland habitats weaved together the various architectural elements on the campus.

Surrounded by waterways, the headquarters building for Giant Interactive Group spans across Zhongkai Road South to create a powerful impression. An expansive green roof of 15,222 square metres envelopes the structure, blurring the edges where landscape and building meet. The rooftop landform culminates in a company guest hotel where private bedroom suites project over the wetland

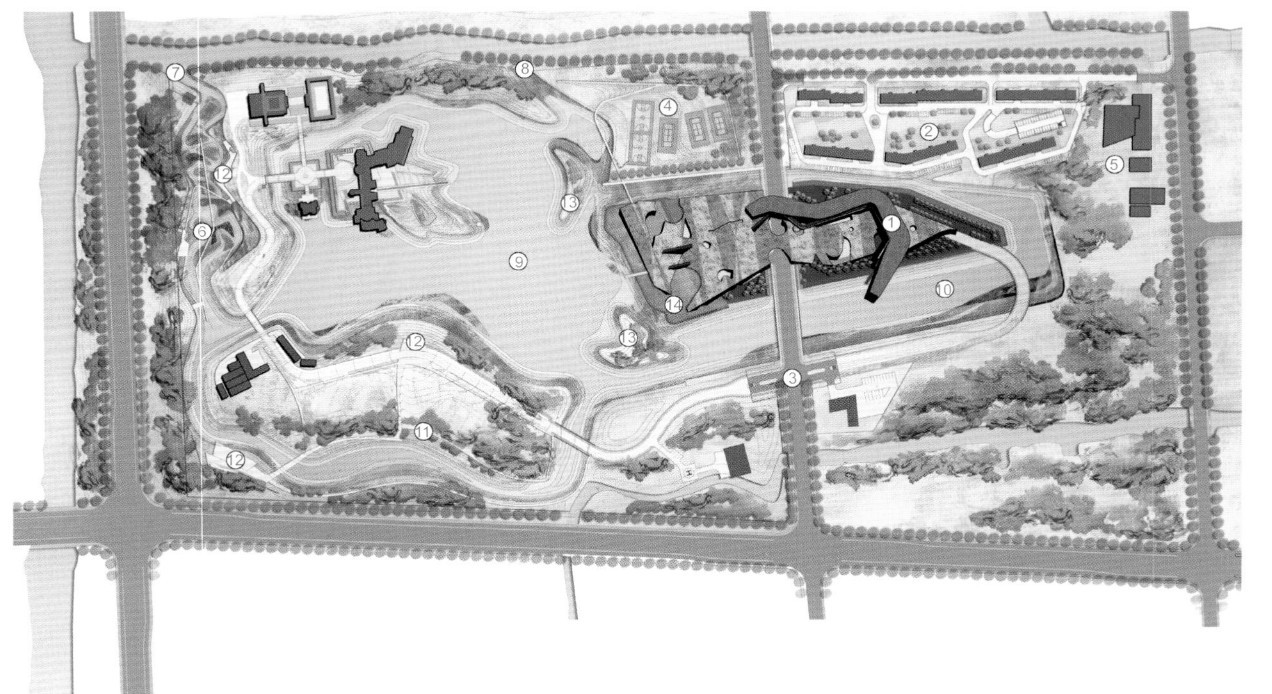

总平面图

1. 总部办公楼
   东侧
   · 不分等级的办公空间
   · 私人办公室
   · 商务套房
   · 图书室
   · 礼堂
   · 展览空间
   · 咖啡厅
   西侧
   · 泳池
   · 多功能体育场
   · 休闲健身空间
   · 宾馆
   （套房下面是湿地池塘）
2. 增建办公室
3. 主入口
4. 休闲运动场
5. 二期开发
6. 湿地公园
7. 水景入口
8. 水景出口
9. 湖泊
10. 水渠（确保水流循环）
11. 小径
12. 观景台
13. 水禽栖息地
14. 湿地池塘

MASTER PLAN

1. Office headquarters
   East end
   •Non-hierarchical office space
   •Private offices
   •Executive suites
   •Library
   •Auditorium
   •Exhibition space
   •Café
   West end
   •Pool
   •Multi-purpose sports court
   •Relaxation and fitness spaces
   •Guest hotel with private bedroom suites overlooking wildlife pond
2. Additional offices
3. Main entry
4. Recreational fields
5. Phase II development
6. Dendritic wetland park
7. Water entry gate
8. Water exit gate
9. Panoramic lake
10. Canal for water circulation
11. Trails
12. Overlooks
13. Waterfowl perch
14. Wildlife pond

EXTENSIVE GREEN ROOF SYSTEM

1. Light planting soil mix
   Ten inches of soil is laid to provide a substrate for plant roofs to grow into as they become established.
2. Filter fabric
   This polypropylene fabric filter helps contain the soil and keep it from washing away, while also acting as a barrier between the planted surface and the building insulation.
3. Stainless steel gabion baskets
   Filled with stone rip rap, the gabions provide structural support and aid in drainage.
4. Drainage board
   Made of strong lightweight plastic, the trays help in the regulation and drainage of rain/irrigation water that percolated through the soil and fabric filter.
5. Insulation
   The insulation works with the vegetation to keep the building interior roughly ten degrees cooler than a standard roof. It also lowers the noise frequency by forty decibels.
6. Root barrier
   Made of polyethylene, this barrier decreases the potential of roots penetrating the concrete structural slab by directing their growth away from the structure of the roof.
7. Waterproof membrane
   This thin sheet composed of rubberised asphalt separates water and drainage runoff from the insulation layer away from the structure of the roof.
8. Concrete cleat
   Reinforced concrete cleats are strategically placed along the roof structure to prevent soil movement across steeply sloped surfaces.
9. Steel angle
   Steel bars attached to reinforced concrete blocks create a frame work to stabilise the soil of steep sloping areas of the roof.
10. Structural slab
    The structural form work which makes up the building's mounded roof.

绿色屋顶结构示意图

1. 轻型混合种植土壤
   厚度：10英寸，约25厘米。为植物的根系生长提供基质。
2. 过滤织物
   一层聚丙烯织物，固定土壤，防止流失，同时在种植表面和建筑隔热层之间起到屏障的作用。
3. 不锈钢笼
   笼内填装石块，提供结构支撑，有助于排水。
4. 排水板
   排水板用坚固的轻型塑料制成，用于从土壤和过滤织物渗透进来的雨水和灌溉水水量的控制和排放。
5. 隔热层
   隔热层利用植被，让建筑室内温度比普通屋顶的建筑低约10度。此外，植被隔热层还能将噪音频率降低40分贝。
6. 根系屏障
   根系屏障层由聚乙烯纤维材料制成，防止植物根系过度生长，穿透混凝土板材结构（屏障层能改变根系生长的方向，不让根系朝屋顶结构的方向生长）。
7. 防水膜
   这层薄膜由橡胶化沥青制成，将水与隔热层和屋顶结构分隔开。
8. 混凝土楔子
   屋顶结构边缘安装了钢筋混凝土楔子，防止土壤由于屋顶坡度过陡而发生移动。
9. 角钢
   角钢安装在钢筋混凝土砌块上，形成稳固的框架结构，让屋顶上陡峭部位的土壤更牢固。
10. 结构板材
    构成屋顶起伏结构的基本框架。

pond. Additional programme is beneath the green roof in the form of a multi-purpose sports court, and fitness areas.

The green roof was designed to be a low maintenance "meadow" that requires little watering and naturalises over time. Unlike a typical green roof, the surfaces fold, soar and dip. The undulating roof structure touches the ground plane, dipping into the adjacent wildlife pond as well as coming into contact with the pedestrian plaza.

The extreme slope conditions vary up to 53 degrees and posed significant challenges for vegetation. An innovative system of reinforced concrete cleats, spanned by steel angles and gabions, are laid parallel to each sloping surface. The system functions as large self-contained cells holding the soil in place and thus minimising slumping and erosion due to gravity. The roof's extensive size acts as a thermal mass that limits heat gain and reduces cooling expenditures.

Because of the roof's folding geometry and orientation, distinct micro-climates occur in response to ridge, valley, sun and shade conditions. After a year-long of testing plants and mix percentages, eleven species were chosen based on the plant's native origin (seven species are native to China), plant rigour, ability to tolerate sun and shade, plant hardiness to dry soil and standing water, and seasonal flowering regimes. The growth rate between plant species is also taken into consideration to prevent species competition. After only one season's growth, the roofscape has become a haven for butterflies, particularly the Small Cabbage White butterfly (Pieris rapae).

Riparian trees and plants, and a wetland sanctuary of networked marshes and islands make a comfortable buffer balancing work and lifestyle. The original programme also includes employee housing set adjacent to the headquarters building and separated by a canal. Several plazas carved from the landscape provide outdoor and recreational spaces for employees, while a central circulation spine and continuous outdoor walkway provide access to the lake and opportunities for gathering. The overall design outcome is the seamless connection of landscape, architecture and environment to the site.

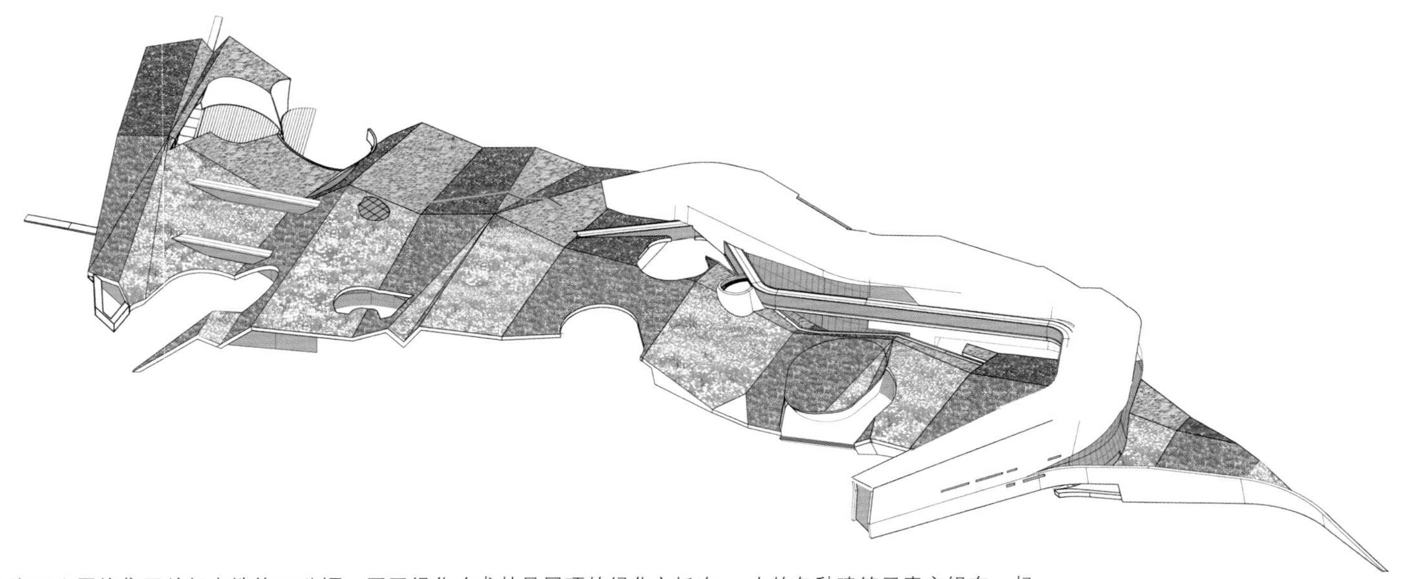

上海巨人网络集团总部占地约18公顷。园区绿化（尤其是屋顶的绿化）旨在打造一座生态公园，一间“活的实验室”。园区内采用开放式布局，建筑与景观紧密结合，不分彼此。园区内一半的面积用作办公用途，另一半面积用于文娱生活，包括宾馆、俱乐部以及餐饮和娱乐设施。

SWA景观设计集团希望景观元素在园区的规划布局中起到框架组织的作用，尤其注重水景的空间体验，打造生态可持续景观。园区用地原本是培育香樟树和桂花树的苗圃，周围是农用河渠。设计师保留了许多原有的树木，在园区内各个位置进行重新栽种。中凯路横贯园区，将土地一分为二。园区的规划利用水元素作为各个部分的连接和组织元素。原有的灌溉水渠、新的蓄水池以及季节性湿地，共同构成错综复杂的园区水系网，打造了多样化的野生生物栖息地的同时，也形成几个风景优美的景点。水景与湿地栖息地将园区内的各种建筑元素交织在一起。

巨人网络集团总部大楼横跨中凯南路，周围遍布各种水景，景色宜人。屋顶绿化面积达15,222平方米，绿色植被覆盖了整座大楼，模糊了建筑与景观之间的界限。屋顶景观的重点是宾馆楼，套房下方就是湿地池塘。绿色屋顶下方还有多种空间，包括多功能体育场和健身区。

绿色屋顶的设计注重未来维护的方便，整体上呈现出一片草坪的形态，不需过多灌溉，日久会呈现出更加自然的形态。本案与一般的屋顶绿化不同，其屋顶不是常规的平面，而是表面曲折蜿蜒，忽高忽低。起伏的屋顶结构时而延伸至地面，时而侵入野生生物池塘，时而与人行广场相交。

PLANTING DIAGRAM　植被组成示意图

Liriope palatyphylla 阔叶山麦冬	Carex oshimensis “Evergold” 金叶苔草	Sedum spectabile “Boreau” 长药八宝	Stachys lanata 绵毛水苏	Vinca major “Variegata” 蔓长春花	Oenothera speciosa 月见草（夜来香）	Tradescantia relexa “Rafin” 水竹草	Spirea x bumalda “Golden Mound” 绣线菊	Ajuga multiflora “Bunge” 多花筋骨草	Herba sedi sarmentosi 垂盆草	Chinese Sedum 中国景天

	Liriope palatyphylla	Carex oshimensis “Evergold”	Sedum spectabile “Boreau”	Stachys lanata	Vinca major “Variegata”	Oenothera speciosa	Tradescantia relexa “Rafin”	Spirea x bumalda “Golden Mound”	Ajuga multiflora “Bunge”	Herba sedi sarmentosi	Chinese Sedum
Group 1 – Ridge 第一组：山脊	5%	30%	5%	25%	5%	10%	20%				
Group 2 – Valley 第二组：峡谷	5%	10%					20%	10%	30%		25%
Group 3 – NE 第三组：东北部	5%	5%			10%	20%	25%			35%	
Group 4 – NW 第四组：西北部			10%		10%	45%		10%			25%
Group 5 – SE 第五组：东南部	5%	5%		25%		35%				25%	
Group 6 – SW 第六组：西南部	5%	30%			10%	20%			25%	10%	
Regular water 正常灌溉											
Moderate water 少量灌溉											
Sun 光照											
Shade 遮阴											

DRAINAGE 排水

● Roof area drain
屋顶排水
----- Direction of roof plane downslope
屋顶平面下坡方向
⟶ Gabion drainage structure
不锈钢石笼排水结构

SLOPE	坡度
0 – 10°	0°~10°
10 – 20°	10°~20°
20 – 30°	20°~30°
30 – 40°	30°~40°
40 – 50°	40°~50°

EXPLOSURE	日光照射
N	北
NE	东北
E	东
SE	东南
S	南
SW	西南
W	西
NW	西北

ON-STRUCTURE
ON-GRADE 屋顶高度

Green roof on-structure
高于地面的绿色屋顶
Green roof on-grade
低至地面的绿色屋顶

Landscape systems diagrams. Variation in drainage, slope, exposure and soil depth contributes to the overall performance of the Giant green roof.

景观设计示意图。排水、坡度、日光照射、土壤深度等方面的多样变化，铸就了巨人集团总部绿色屋顶的整体效果。

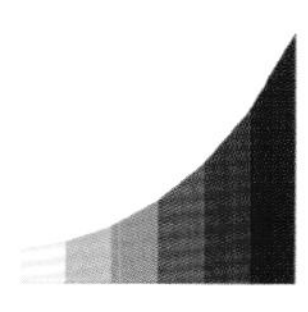

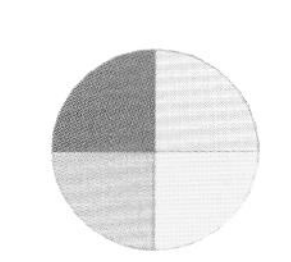

屋顶坡度很大，最大的地方有53度，这给绿化带来了很大挑战。设计师创新采用钢筋混凝土夹板，用角钢和生态格网加以固定，夹板与屋顶平行设置。夹板内自带单元格状的土壤层，能够有效防止土壤在重力作用下发生滑动、坍塌和侵蚀。巨大的屋顶起到隔热层的作用，减少了太阳能的吸收，降低了空调制冷的费用。

由于屋顶折叠式的形态和朝向，在屋脊、谷地、阳光和阴凉等条件的共同作用下，屋顶上形成了独特的微气候。在为期一年的测试之后（包括植物种类及其搭配的百分比），设计师最终选取了11种植物。植物的选择考虑了原产地（其中7种原产中国）、植物生命力、喜阳或喜阴的习性、对干旱土壤和积滞水的耐受力以及开花的情况等。不同植物种类的生长速度也列入考虑范围，防止品种恶性竞争。一季度的生长之后，屋顶已经成为蝴蝶的天堂，尤其是菜粉蝶。

滨水树木和植物，包括由沼泽和小岛构成的湿地景观，营造出舒适美观的环境，平衡了工作与生活的差距。总部大楼旁边还有员工宿舍，二者之间有一条水渠。景观元素划分出若干个小广场，为员工提供了户外休闲空间。园区的中央大道以及连绵的散步道四通八达，通向湖边，通向各个小景点。园区整体设计实现了景观、建筑与环境三者的“无缝对接”。

Great Ormond Street Hospital Rooftop Garden

大奥蒙德街医院屋顶花园

Location:
London, UK
Architect:
Spacelab
Landscape architect:
Andy Sturgeon Landscape Design
Client:
Great Ormond Street Hospital
Photographer:
Rama Knight
Area:
280 sqm

项目地点：
英国，伦敦
建筑设计：
“空间实验室”建筑事务所
景观设计：
安迪·斯特金景观设计公司
客户：
大奥蒙德街医院
摄影师：
哈玛·奈特
面积：
280平方米

Project description:

The "Friends Garden" that is funded by the Charity "The Friends of Great Ormond Street" represents much more than just a pavilion and garden to the Charity and the hospital staff. The Friends Garden has become, as it was always intended to be, an escape into an environment that refreshes and replenishes after the rewarding but quite stressful work that the staff of the hospital carry out on a day-to-day basis.

The Friends Garden is the "Friends of Great Ormond Street" charity way of saying in a positive way a thank you to the staff for all their hard work by rewarding them with a unique and fantastic space which they can use to relax and enjoy life. Spacelab and Andy Sturgeon Garden Design have created, on a very restrained site, a space that can be enjoyed throughout the year during the day or night. The hospital and staff regularly uses the space for events, and staff can book out the space for private functions such as birthdays and receptions.

As well as providing a great space for the staff to relax it also provides people with a space to contemplate and remember the two members of staff of Great Ormond Street Hospital who tragically died as a result of the 7/7 bombings in London. The hospital approached the families and friends of the two members of staff who provided words and phrases that they believed best represented them and these have been strategically placed around the garden providing a commemorative memorial which is absorbed into the garden rather than being a focus.

METRO

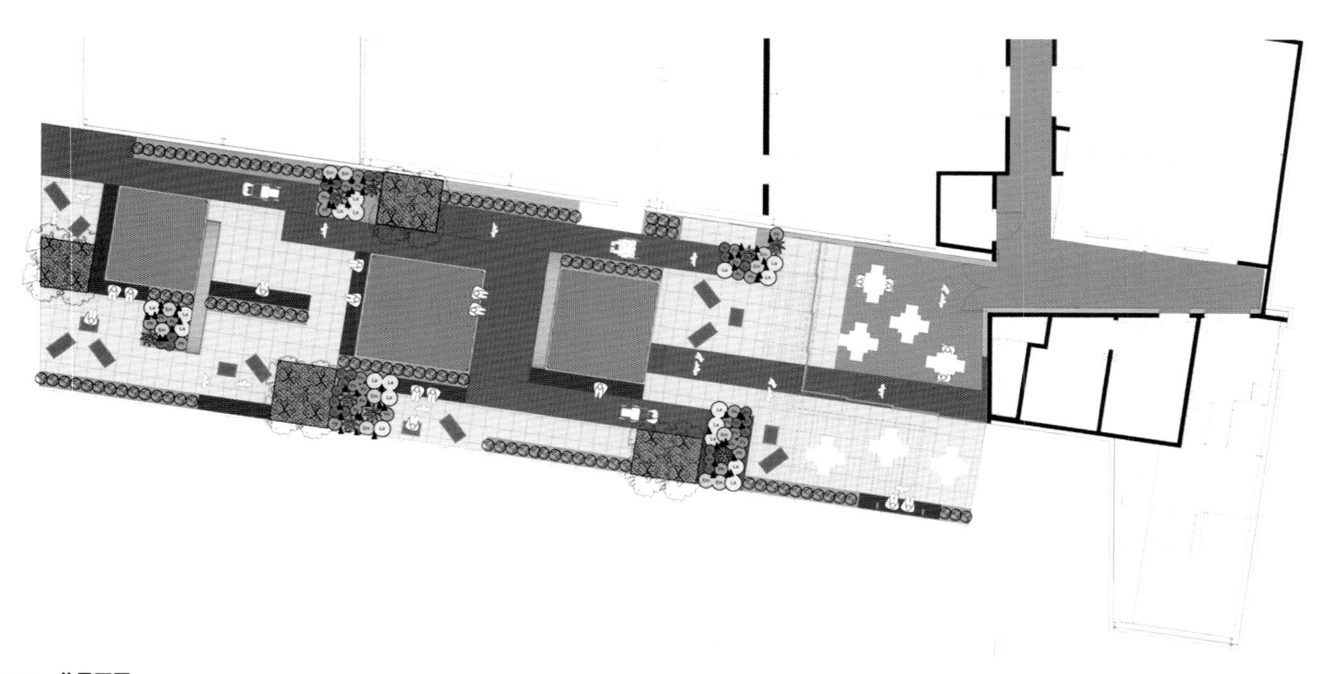

SITE LAYOUT 总平面图

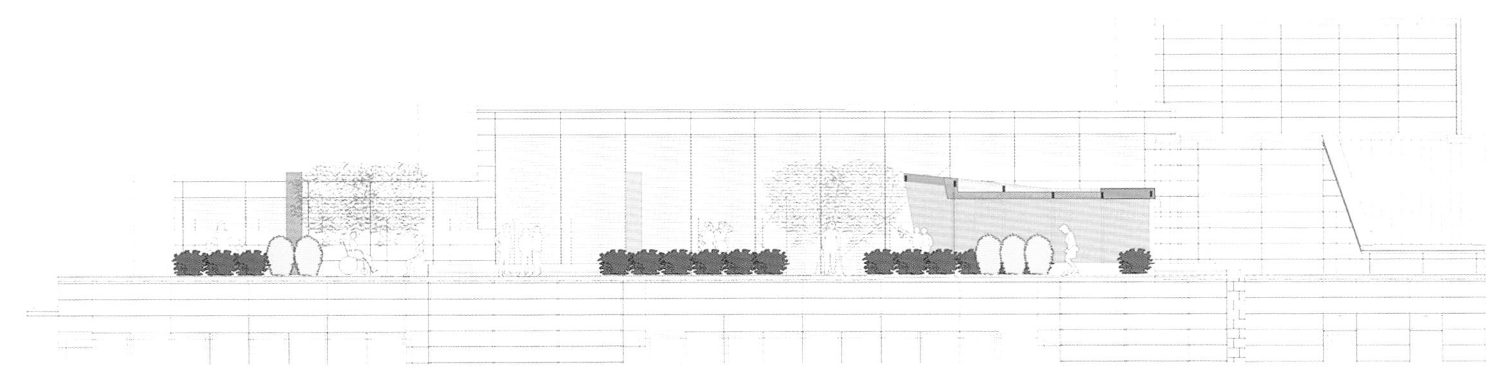

SITE ELEVATION 总立面图

Materials:

Pavilion

- Aluminium framed glass sliding / folding doors
- Aluminium polyester powder-coated cladding
- Anodised aluminium louvres

Terracing

- Hardwood decking
- Stone paving
- Polyester powder-coated planters
- Glulam timber beams / hoops
- Hardwood seating / benching

“友谊花园”是“大奥蒙德街友谊慈善会”资助兴建的一座屋顶花园。它不仅仅是为慈善会和医院员工增加一点休闲设施。医院的工作虽然意义非凡，但是日复一日的紧张工作也让人压力巨大。“友谊花园”已经成为医院员工在一天繁忙的工作之后放松身心，为自己加油充电的好地方。

“友谊花园”是“大奥蒙德街友谊慈善会”以自己独特的方式对医院员工表示感谢，用一个世外桃源的环境来感谢他们辛苦工作。员工可以在这里休闲娱乐，享受生活。“空间实验室”建筑事务所和安迪·斯特金景观设计公司携手，虽然场地条件非常有限，但还是打造出一年四季不论白天还是夜间都能供员工使用的绿色空间。医院可以在这里举办常规活动，员工也可以预约场地举行私人聚会，比如生日聚会或者接待活动。

除了为员工的休闲生活提供场地之外，“友谊花园”还有一项功能，那就是作为一个沉思冥想的空间，让人们怀念大奥蒙德街医院在伦敦“七七爆炸案”中不幸丧生的两名员工。医院找到这两名员工的家人和朋友，从他们那里得到最能表述两人品德的词句，然后将这些词句刻在花园的各个角落。没有突兀的纪念碑，相反，纪念碑已经融入花园本身；整个花园就是一座独特的纪念碑。

LOCATION PLAN 位置图

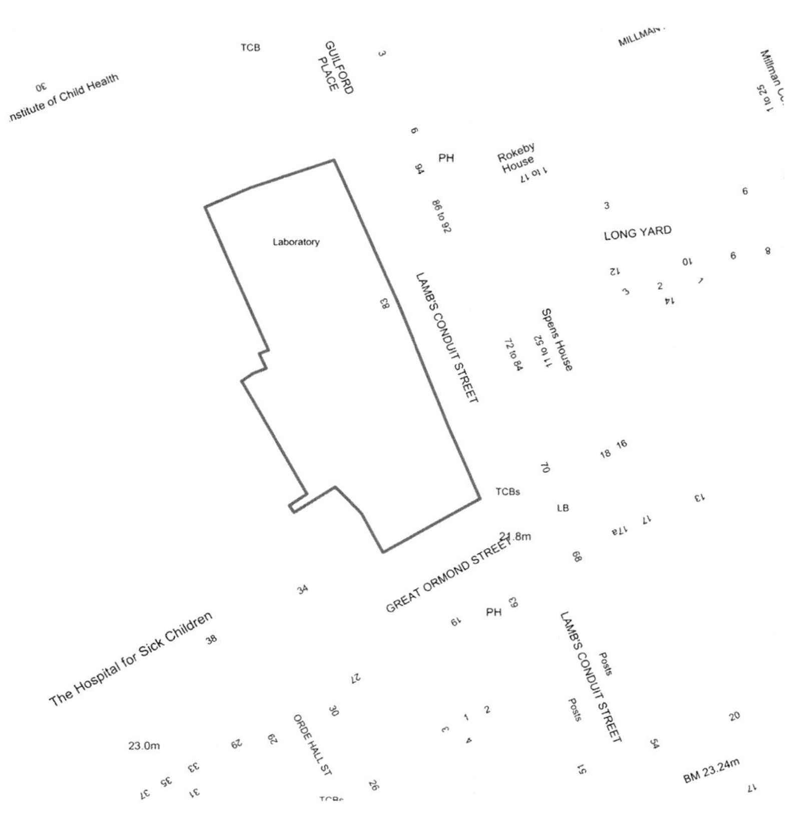

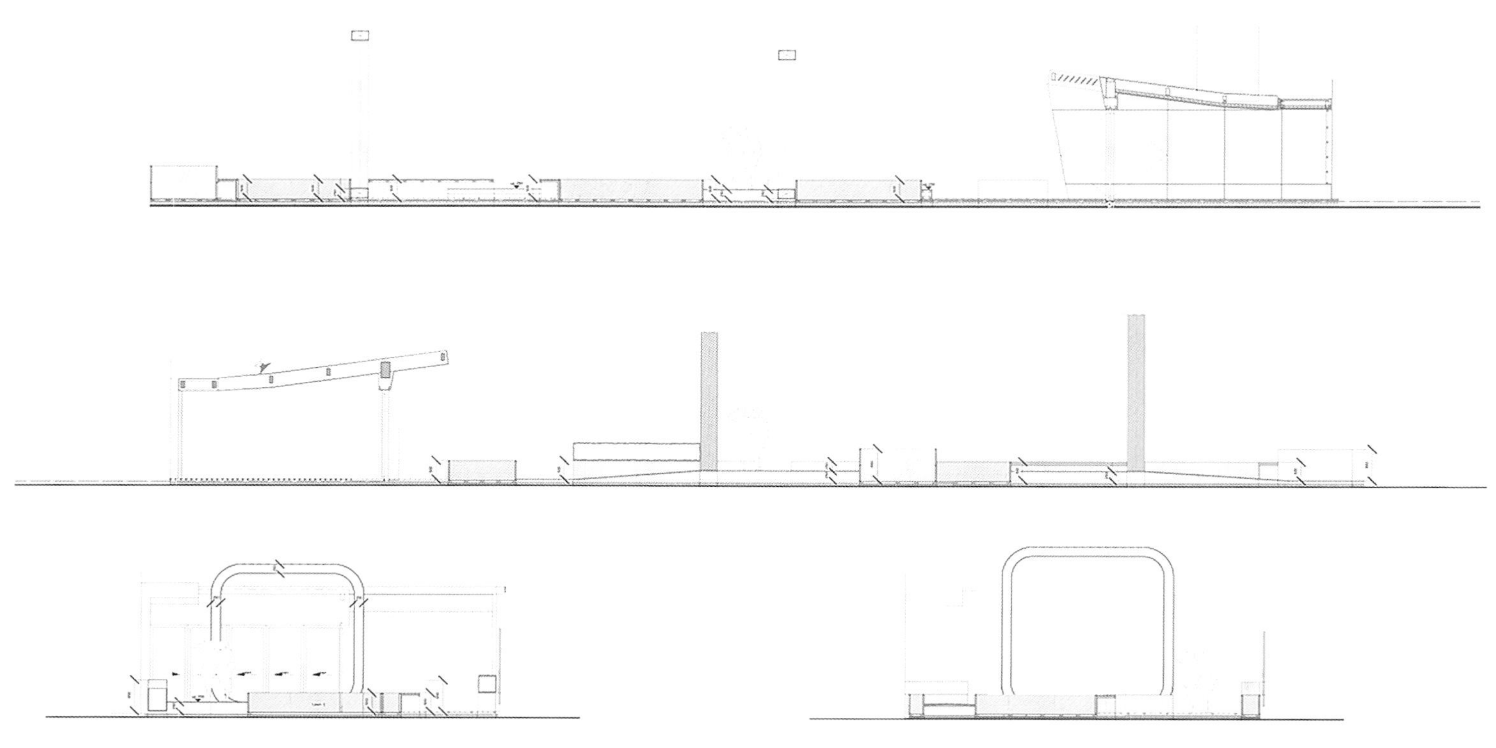

SITE SECTIONS 总剖面图

所用材料包括：

凉亭结构：

- 铝制框架的玻璃拉门/折叠门
- 铝制聚酯电镀膜
- 电镀铝遮光栅格

平台：

- 硬木地板
- 石材铺装
- 铝制聚酯电镀种植槽
- 胶合木横梁/圆环
- 硬木座椅/长椅

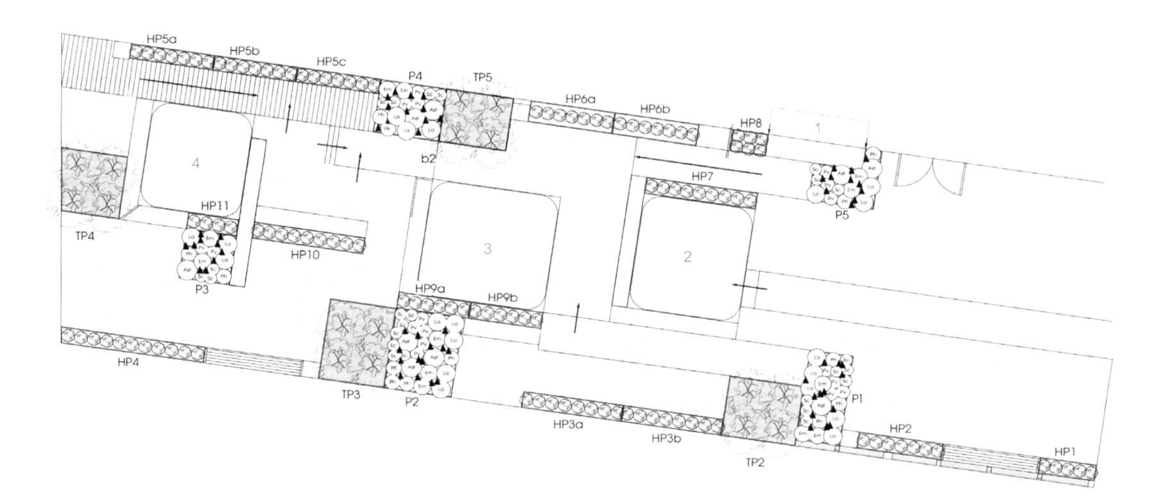

PLANTING PLAN

1. Vent area
2. Lawn 1
3. Lawn 2
4. Lawn 3

植被平面图

1. 通风区
2. 草坪1
3. 草坪2
4. 草坪3

KEY

P: Planter

HP: Hedge Planter

TP: Tree Planter

注释：

P：种植槽

HP：树篱

TP：树木

PLANTING KEY 植被示意图

X	16 No. Carpinus betulus “Frans Fontaine” 20-25cm girth, 1.8m clear stem 欧洲鹅耳枥（围长20厘米～25厘米，茎长1.8米）		12 No. Euphorbia x martini 5L pots @ 500mm centres 美洲锦地草
	Trees underplanted with 60 No. Miscanthus “Yakushima Dwarf” 2L pots @ 450mm centres 大型树木下种植芒草	La	19 No. Lavandula angustifolia 10L pots @ 550mm centres 狭叶薰衣草
	114 No. Taxus baccata hedging 1,000mm high (cut back to 800mm high) @ 2.5 plants/linear metre 欧洲红豆杉树篱（生长高度1,000毫米，修剪为800毫米）		13 No. Pennisetum alopecuroides “Hameln” 10L pots @ 450mm centres 狼尾草
Pv	17 No. Panicum virgatum “Heavy Metal” 5L pots @ 400mm centres 柳枝稷		24 No. Salvia nemorosa “Caradonna” 5L pots @ 350mm centres 林地鼠尾草
Ast	9 No. Astelia chathamica “Silver Spear” 10L pots @ 600mm centres “银矛”百合		60 No. Tulipa “Ballerina” “芭蕾”郁金香
			Total area approx. $36m^2$ 草坪总面积约为36平方米

Midtown Manhattan Sky Garden

曼哈顿市中心空中花园

Completion date:
2011
Location:
New York, USA
Architect:
Gertler & Wente Architects
Landscape architect:
HMWhite
Client:
Shorenstein Realty
Area:
604 sqm

竣工时间：
2011年
项目地点：
美国，纽约
建筑设计：
格特勒&温特建筑事务所
景观设计：
HMWhite景观事务所
客户：
舒思深房地产公司
面积：
604平方米

Project description:

Who would expect to experience an undulating wildflower meadow suspended 17 floors above Midtown Manhattan within a mid-century Emery Roth office building? Originally a barren and inaccessible wrap-around terrace, braided meadow grass and seasonal flowering perennials and bulbs are now the immediate eye-catching foreground to Shorenstein Realty's offices and conference rooms. Staff feedback on the impact of the garden's physical and visual proximity in their daily work experience includes measurable improvements in productivity, creative thinking and overall sense of well-being. While the garden design's rich and dynamic canvas adds significantly to the building's value, the return on this investment with increases in workers' performance becomes inexhaustible annuity.

Redefining the 17th floor of the historic Western Publishing Building in midtown Manhattan, HMWhite created a contemporary garden terrace as an extension of Shorenstein Properties' new corporate offices. The 604-square-metre living roof establishes a biophilic transformation of the office environment and daily employee experience. A lush garden wraps the building and visually merges interior office spaces with a meadow landscape.

The entire garden is carefully engineered to knit with building infrastructure to prevent potential wind uplift, accommodate significant weight restrictions and support a sophisticated, multi-seasonal plant palette tolerant of extreme growing conditions. The elongated meadow design incorporates drought-tolerant

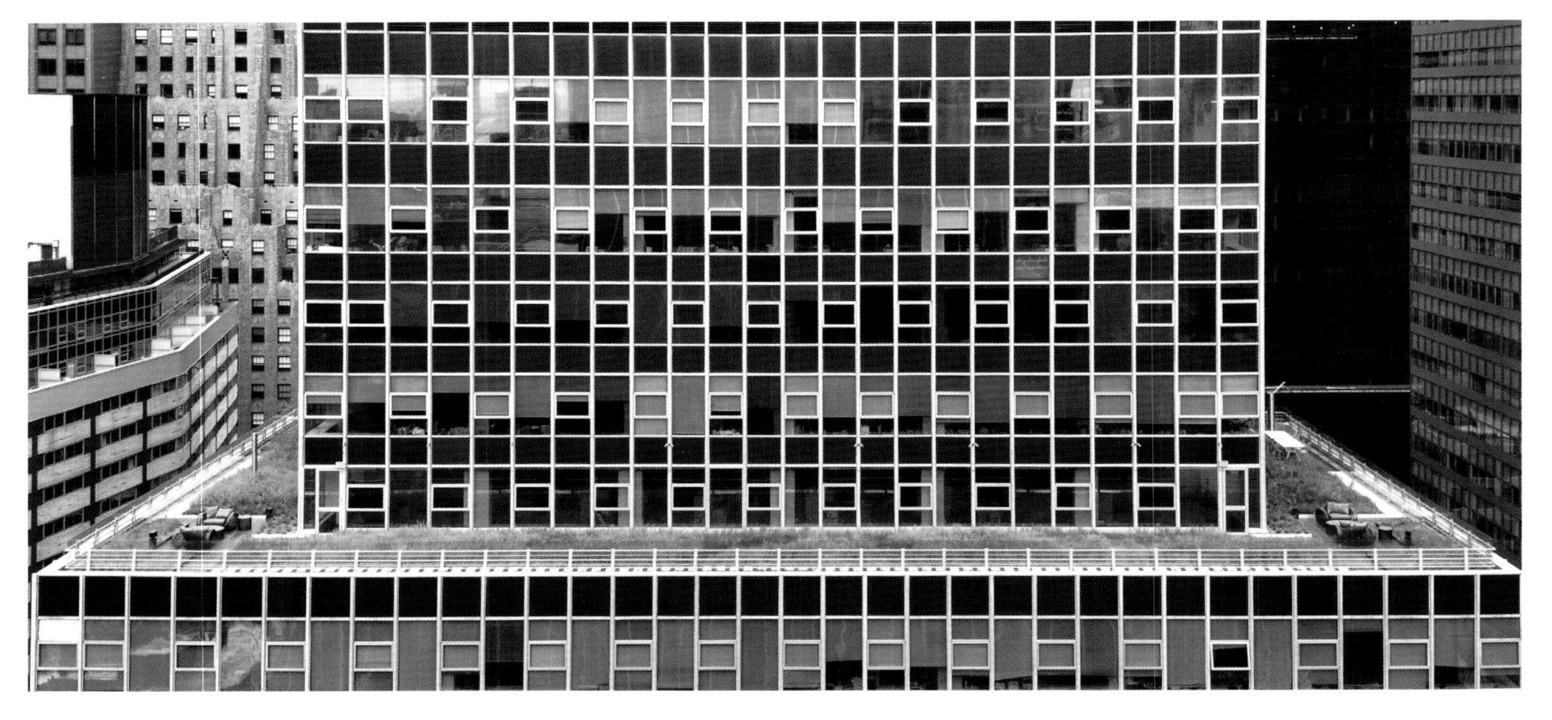

grasses, perennials and spring bulbs – punctuated by flowering trees in structurally specific locations.

The office interior links with the exterior via a series of wooden platforms embedded in the landscape creating flexible "break-out" meeting spaces and places of respite. The intimate garden rooms connect with a perimeter path system framed by a translucent wind protection screen along the building's edge.

On the northwestern side of the building, an isolated roof space is transformed into an outdoor golf putting green providing a retreat instantly popular with staff.

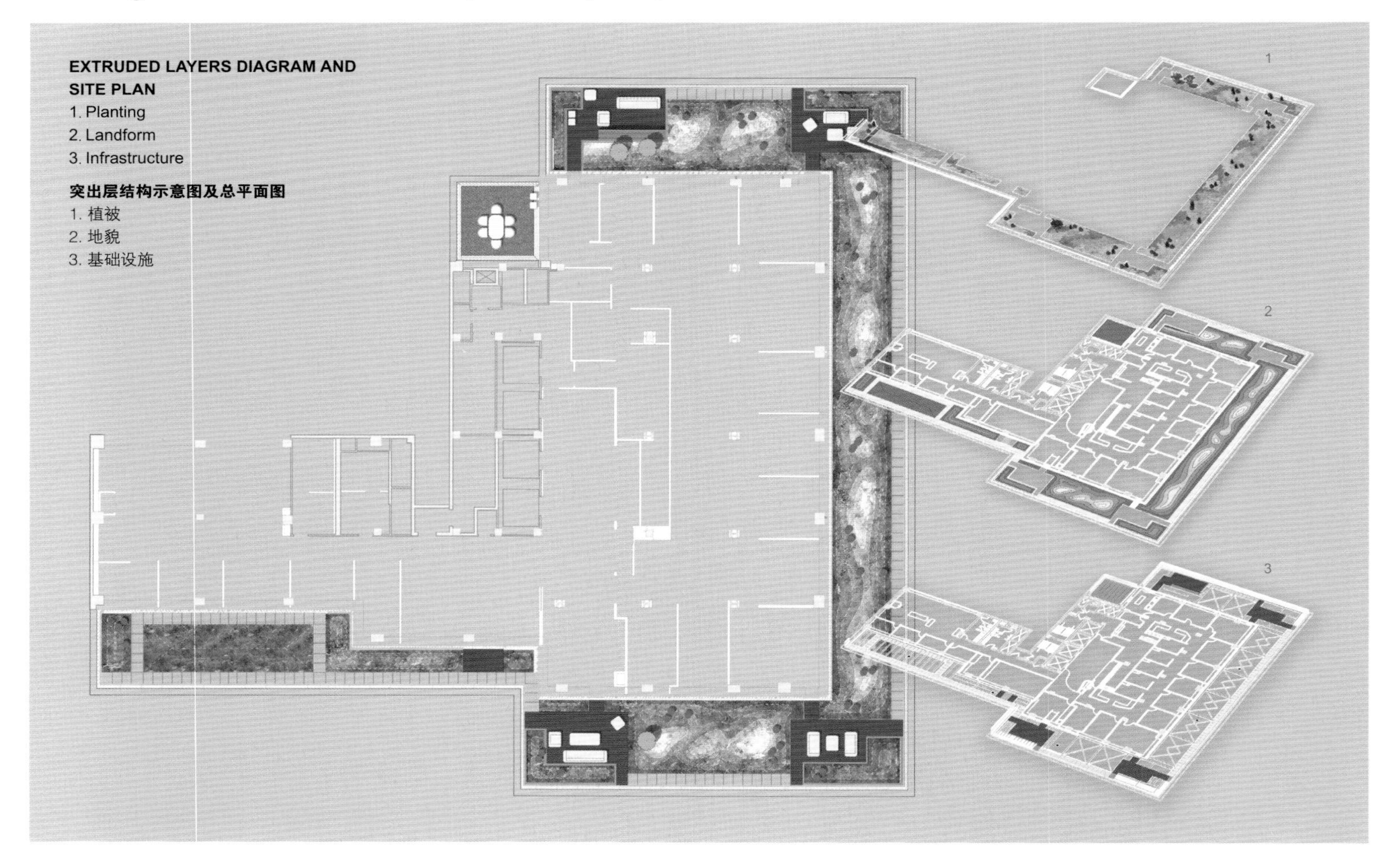

EXTRUDED LAYERS DIAGRAM AND SITE PLAN
1. Planting
2. Landform
3. Infrastructure

突出层结构示意图及总平面图
1. 植被
2. 地貌
3. 基础设施

NORTHWEST SECTION　西北剖面图

SOUTHWEST SECTION　西南剖面图

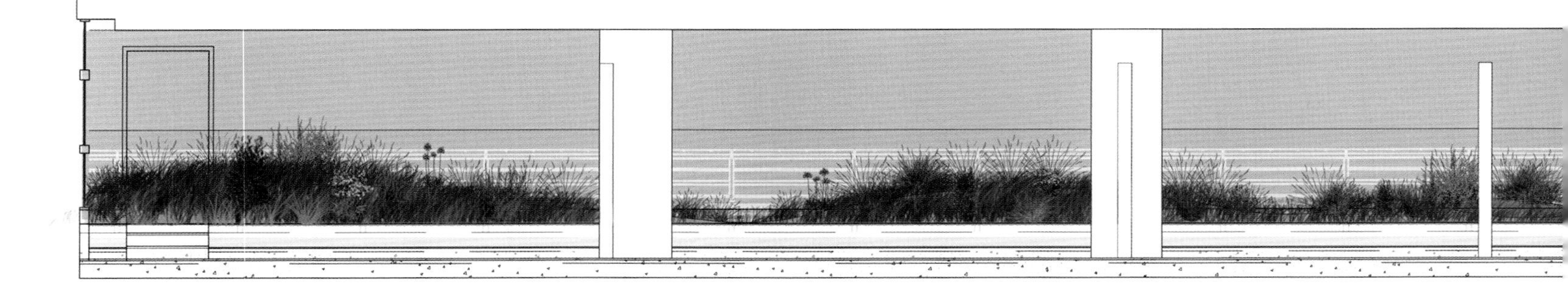

谁能想象在曼哈顿市中心由建筑大师埃默里·罗斯设计的中世纪办公楼17楼的高度上建一座空中花园？这里原来只是围绕着办公楼的一圈光秃秃的平台，无法供人使用；而现在，绿油油的草皮和随季节盛开的鲜花让平台变身为一座极其吸引眼球的空中花园，成为舒思深房地产公司办公室和会议室的美丽风景。公司员工对这座花园的反馈是：工作绩效显著提高，有助于创造性思维和身心愉悦。空中花园的设计让这栋大楼的价值显著增加，不仅如此，它带来的员工工作表现上的进步还将产生无穷的价值。

这栋建筑历史悠久，原来是一栋出版大楼，坐落在曼哈顿市中心。HMWhite景观事务所在其17层的平台上打造了一座现代风格的空中花园，作为舒思深房地产公司新办公楼的延伸空间。花园面积约604平方米，在单调的办公环境和美丽的自然风景之间建立了一种过渡。花园上植物生长非常茂盛，仿佛给大楼围上一条绿色的腰带，花园上的美景自然地融入到室内办公空间中。

整个花园的结构设计非常谨慎，与建筑结构紧密契合，避免植被被风掀起。此外，还要确保平台能够支撑植被和土壤的重量。植物品种的选择也有讲究，设计师精心选取适合各个季节生长的植物，并且要能够适应这个高度上严酷的生长环境。草坪呈狭长形，种植的都是耐旱的草本植物，个别位置上还种了几棵树。

嵌入草坪中的一系列木质平台将室内办公空间与室外景观联系起来，这里可以灵活地用作非正式会议空间或者休闲区。平台上划分出多个小花园，由一条小径连接，小径旁边沿着建筑物外围安装了半透明的防风屏风。

建筑西北侧专门开辟出一个户外迷你高尔夫球场，广受员工欢迎。

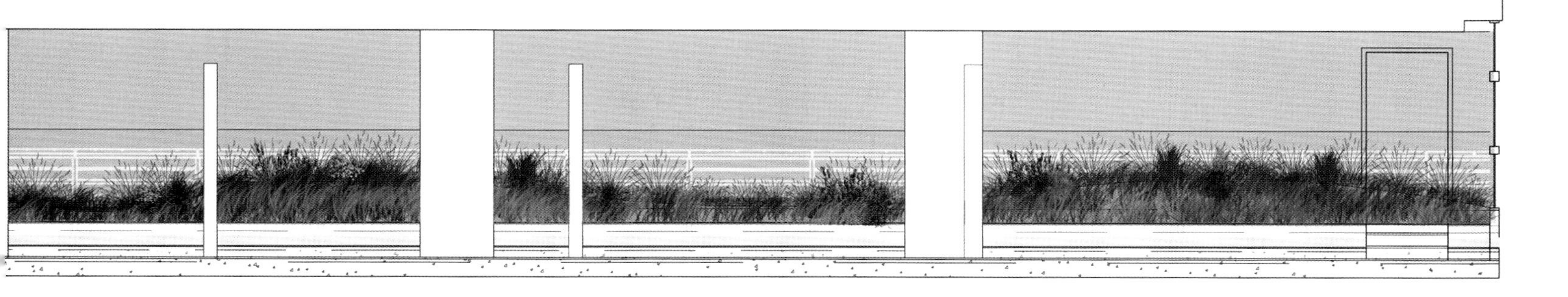

ILLUSTRATIVE ELEVATIONS THROUGH WINDOWS
立面示意图（透过窗口）

MIMA Mixed-use Residential Green Roof
米玛多功能住宅楼屋顶绿化

Completion date:
2011
Location:
New York City, New York, USA
Architect:
Arquitectonica
Landscape architect:
HMWhite
Client:
Related Company
Photographer:
HMWhite
Area:
1,336 sqm
Awards:
2013 ASLA New York Chapter Honour Award

竣工时间：
2011年
项目地点：
美国，纽约州，纽约市
建筑设计：
美国ARQ建筑设计事务所
景观设计：
HMWhite景观事务所
客户：
美国联营房地产公司
摄影师：
HMWhite景观事务所
面积：
1,336平方米
奖项：
2013年美国景观设计师协会纽约分会荣誉奖

Project description:

Emerging from the third floor of the Related Company's LEED® Gold Certified 63-storey complex near Times Square, HMWhite designed a pair of unique outdoor garden terraces hovering over the bustling city streets. The terraces link building interior amenity spaces with a series of outdoor garden rooms configured for a variety of uses by the building's residents.

A long linear space overlooking 42nd Sstreet, the north terrace wraps the eastern end of the tower. A contiguous meadow landform punctuated with groves of columnar Hornbeams unifies and shapes a series of social pods. This common landscape brings human scale to each pod, designed as flexible spaces for day and evening activities.

The south terrace's design modulates divergent structural slab elevations to create a barrier-free outdoor room extension of the lap pool and fitness facility. Linear steel planters weave through the terrace to emphasise shifting surface elevations and support seasonal blooming shrubs and perennials. A central sloping lawn reflects the interior pool and is flanked by elevated sunning decks that transform into an outdoor movie theatre and event space in the evening.

DETAILED PROJECT FEATURES
PROJECT PURPOSE:
Establish amenity-rich roof terraces to ser-

vice residents from the 700 rental apartments and the 150 tower residences between two separate terraces – an 836sqm north terrace and a 500sqm south terrace. Intent was to programme and design a series of outdoor garden rooms that offer a diverse variety of social gathering functions as well as private party spaces and dining alcoves as extensions of the building's amenity lounge and fitness club floor uses. The success of the terraces' landscape design was considered essential to distinguish this residential development during a highly competitive and down-market period. The terraces' innovative green roof designs, complete with their thermal battery protection, stormwater absorption and integration of drought-tolerant native plantings, were significant contributors to the project's Gold LEED certification goal. The obvious additional by-product of the terraces' rich and seasonal plantings

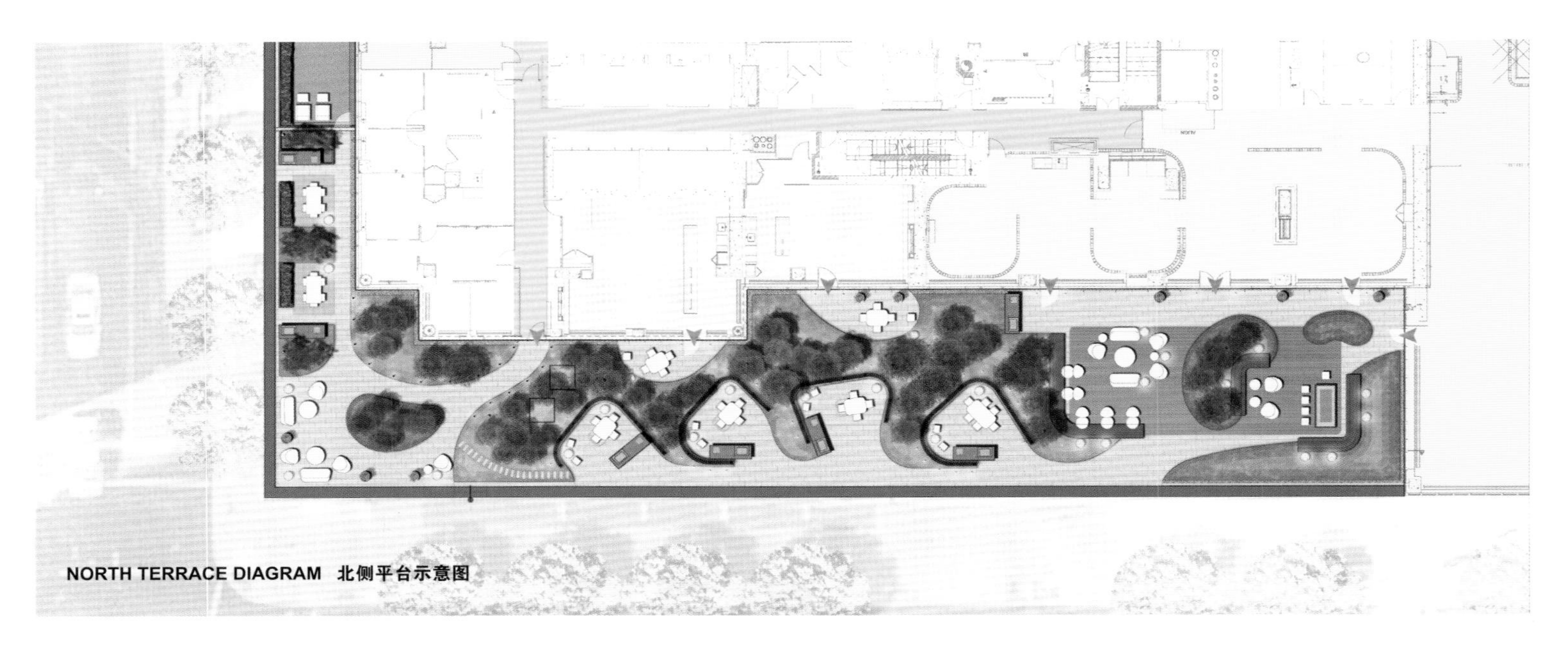
NORTH TERRACE DIAGRAM 北侧平台示意图

米玛多功能住宅楼离纽约时代广场不远，是美国联营房地产公司（Related Company）开发的一栋60层的超高建筑，获得了美国绿色建筑协会LEED金级认证。HMWhite景观事务所在这栋建筑的三楼设计了两个户外平台，使之成为悬于繁华大街之上的两个空中花园。绿色平台为建筑的室内空间拓展出户外花园，花园里有多种多样的休闲设施供住户使用。

北侧平台是一个狭长的绿色空间，下方是42号大街。这个平台的绿化采用连绵的草地，间或种植一些树木（鹅耳枥），使整个平台呈现出统一的景观风格，同时界定出一系列休闲区。适当的景观设计使休闲区看上去具有人性化的体量，空间设计很灵活，适合白天或夜间进行各种活动。

南侧平台由于建筑结构的原因，地面高差较大。设计师克服了这个难题，将其打造成无障碍的户外休闲空间，是小游泳池和健身中心的延伸。线性钢质

SUSTAINABLE DESIGN STRATEGIES:

- LEED® Gold Certified
- Solar-reflective concrete pavers and certified hardwoods
- 5,500 gal rain water cistern recycled for irrigation
- Water-conserving Netafin irrigation system
- Drought-tolerant native plant selections
- Low landscape management requirements

可持续设计策略：

- LEED金级认证
- 地面铺装采用能够反射太阳光的混凝土材料和硬木
- 蓄水池能存储20,820升雨水，用于灌溉
- 节水的灌溉系统
- 选用耐旱的本地植物
- 景观维护需求不高

provide a transformative contrast to the building's Times Square bustling commercial district character.

LANDSCAPE ARCHITECT'S ROLE:
HMWhite led the roof garden terrace design vision, programming, design development, construction documentation and administration, which included all surface materials, green roof and planting infrastructure, outdoor kitchen designs and equipment, custom-designed seating and planters, layout and furnishings selection. HMWhite established the supporting lighting design concept that included fixture selections, their final layout and coordination with the lighting design consultant. Other team design and technical coordination included roof terrace components with building architectural design, structure and utilities, outdoor theatre audio and visual components, and subtle integration of security cameras. HMWhite also designed the building's surrounding streetscapes and entrance, featuring a family of custom-designed and -fabricated iconic planters. Post construction landscape management practices were established by the design team, its arborist and food web soil specialist to ensure vital sustainable growing conditions that uphold landscape design intentions.

SITE CONTEXT:
Emerging from the third floor of a sheer glass 63-storey tower overlooking West 42nd Street, the north terrace offers an oasis and soothing escape from its Times Square neighbourhood while providing a variety of spectacular Midtown Manhattan skyline-framed views. The south terrace's sunny exposure and transparent link to the fitness centre's lap pool presented an ideal sun terrace design opportunity. The theatre's four-storey fly tower presented a blank eastern wall to evolve into an ideal outdoor theatre screen. The garden's opposite side features a grove of tall growing Aspens trees to screen and buffer a future MTA utility building rising three storeys above the terrace.

SPECIAL FACTORS:
SITE DESIGN:
NORTH TERRACE
Continuous undulating meadow landforms, punctuated with groves of tall columnar Hornbeams, unify and give shape to

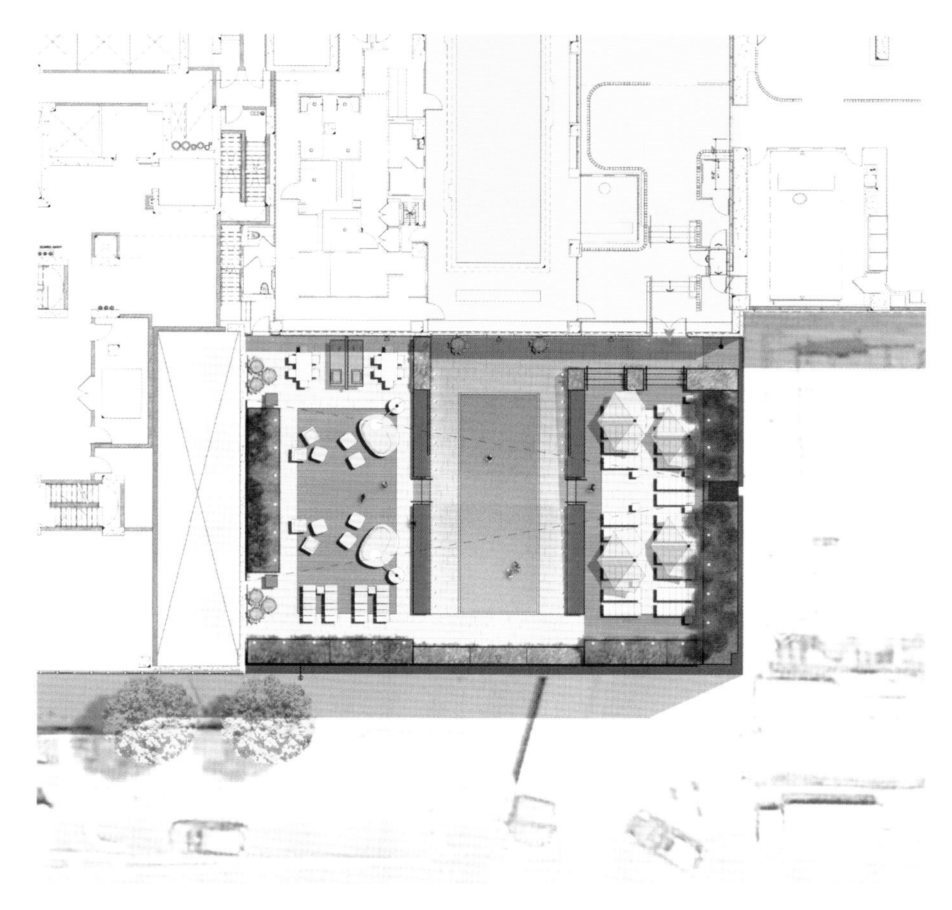

SOUTH TERRACE DIAGRAM

南侧平台示意图

a series of distinct social pods proportioned to break down the monotony of the terrace's 97.5-metre-long linear space. The subtle emergence of the mounded common landscape provides a "false-ground" appearance, further reinforced by the diverse seasonal blooming and textural horticultural events. Double-sided vine screens create spatially efficient, gardenesque private grilling and dining rooms. They also offer another layering garden element from the private party room break-out areas and the larger club and public lounges. A 74sqm dog park adjacent to the building's Dog Spa is protected and defined by the continuation of the terrace's ribbed timber screens and hedge plantings – a welcome respite for both owner and pet.

SOUTH TERRACE

The south terrace's design modulates divergent structural slab elevations to create a barrier-free outdoor room extension to the fitness centre and lap pool. A central sloping lawn reflects the interior pool and is flanked by elevated sunning decks that transform into a popular outdoor movie theatre and event space in the evening. Distinct types of fixed and moveable lounging furnishings accommodate a variety of planned and spontaneous social gatherings. An outdoor shower, grilling stations and dining tables have been seamlessly nestled into the garden terrace's native plantings and minimal material palette.

SHARED DESIGN FEATURES:

Planting Design: Minimal and bold palette of fine textured low grasses with spring flowering bulbs contrasted with coarse textured flowering

种植槽是贯穿整个平台的衔接元素，里面种植着随季节开花的灌木和草本植物，将变化的地面高度连接起来。中央是一块斜坡草坪，与室内的游泳池相互呼应，草坪旁边是架高的日光平台，夜晚可以用作户外影院，或者举行各种活动。

设计目标：

打造休闲设施齐全的屋顶平台，北侧平台面积为836平方米，南侧平台面积为500平方米，共同为700户出租公寓住户和150户常住居民提供休闲空间。设计目标是规划并设计一系列户外绿色空间，满足居民社交活动的各种需求，同时包含私密的聚会空间和用餐空间，用于室内空间的拓展（这一楼层的室内空间是健身俱乐部）。平台景观的成功设计让这一楼盘在激烈的竞争中，尤其在房地产市场低迷的时期，能够脱颖而出。平台上极具创意的空中花园，再加上蓄能电池、雨水处理、种植耐旱的本地植物等设计，让本案得以申报LEED金级认证。平台上茂盛的植物还有一项附加价值，那就是让这栋建筑与时代广场这样商业气息浓郁的街区形成鲜明的对比。

景观设计师扮演的角色：

HMWhite景观事务所全权负责本案平台花园的蓝图规划、设计开发、施工监理等环节，包括材料的选择、绿化基础设施的设计、户外厨房设计和设备选择、专门设计的坐具和花坛、空间布局以及户外陈设的选择等。HMWhite景观事务所还提供了相应的照明设计，包括灯具的选择和最终布局，并与灯光设计顾问紧密合作。其他设计和技术工作还包括屋顶平台组件的设计（与建筑结构紧密结合）、户外影院的视听设备以及监控摄像头的巧妙安装。HMWhite景观事务所还设计了大楼入口和周围的街道景观，尤其是亲自设计并装配了特色花坛。设计团队对工程竣工后的景观维护工作也作了规

perennials as spatially organizing and unifying garden design techniques.

Garden Materials: Introduced certified IPE decking and custom-designed benches as an additional built organic material to add warmth to offset the building's glass and steel expanse.

Lighting Design: Unique floor lamps, pencil-thin path lights and up-lights washing the texture surfaces of the Ivy-clad vine screen walls and the Hornbeam canopies.

Project Economics: Provided cost-estimating and value-engineering strategies through each design and documentation phase to ensure strict project budget adherence. Sourced and worked closely with local fabricators for steel edging and custom planter fabrications.

划，由树木栽培专家和食物链专家亲自把关，确保屋顶花园维持良好的生态条件，让植物能够长久生长。

场地条件：
这是一栋63层的超高建筑物，采用玻璃外立面，平台位于三层。北侧平台是一块静谧的绿洲，下方是繁华的西42号大街。它脱离了时代广场的商业气息，拥有眺望曼哈顿市中心天际线的绝好视野。南侧平台采光很好，与健身中心的小泳池相连，是设计日光平台的最好场地了。东侧有一面墙壁，正好可以用作户外影院的屏幕。户外影院的另一边种植着高大的山杨树，形成一道屏障，起到保护作用（未来这一侧会修建起比平台高三层的建筑物）。

场地设计：
北侧平台
北侧平台的设计采用连绵的草地，间或种些高大的鹅耳枥树，让一系列独特的休闲区掩映在统一的景观背景下，同时这些休闲区也将长约97.5米的狭长平台切割成人性化的小空间。平台上的景观设计非常丰富，让人产生身在地面的错觉，尤其是当各种植物随季节生长、开花，呈现出一派生机的时候。藤蔓屏障将空间划分出烧烤区和用餐区，全都是花园一般的环境。对于私人聚会空间的休闲区和宽敞的俱乐部公共休闲区来说，藤蔓也是一个园艺元素。宠物水疗区的旁边是74平方米的“宠物公园”，平台上连绵的木栅栏和树篱起到保护的作用。这里是宠物狗和它们的主人都十分喜爱的地方。

南侧平台
南侧平台的设计克服了地面高差的难题，打造了无障碍的户外绿色空间，是健身中心和小型泳池的延伸。中央的坡地草坪与室内泳池相互呼应，旁边是日光平台，夜晚可以用作户外影院，也可以举办各种活动。坐具等家具陈设，有些是固定的，有些是可以灵活布置的，全都别具一格，能够满足居民各种社交活动的需要。户外淋浴间、烧烤台以及餐桌都与平台上栽种的本地植物融为一体。选用的材料种类不多，营造出简单清新的环境。

南北平台的设计共同点
植被设计：草坪选用的草本植物种类不多，但效果很好，极具视觉冲击性，此外还有春季开花的鳞茎类植物以及比较粗糙的多年生开花植物。植被是进行空间布局和统一风格的重要设计元素。

材料选用：平台地面和专门设计的长椅采用重蚁木。有机材料的使用为平台增添了一丝温暖的气息，消解了这栋大楼的玻璃和钢材外立面带来的冰冷感觉。

照明设计：独特的地灯、纤细的小径照明灯以及专门为爬满常春藤的分隔墙和鹅耳枥树的树冠照明的向上照射的灯。

成本效益：在设计过程中的每个阶段都估算成本和价值，确保严格按照预算来进行。钢材切割和特制花坛的制作都与本地制造商合作，降低成本。

One & Ortakoy

奥塔科伊综合楼

Completion date:
2011
Location:
Istanbul, Turkey
Landscape Project:
DS Architecture – Landscape
Landscape Architect:
Deniz Aslan
Architect:
GAD Architecture
Client :
Doğu İnşaat A.Ş
Photographer:
Gürkan Akay
Area:
3,000 sqm

竣工时间：
2011年
项目地点：
土耳其，伊斯坦布尔
景观公司：
DS建筑事务所
景观建筑师：
德尼兹・阿斯兰
建筑设计：
GAD建筑事务所
客户：
Doğu İnşaat公司
摄影师：
戈尔坎・阿凯
面积：
3,000平方米

Project description:

One & Ortakoy is a mixed-use complex in the Istanbul neighbourhood of Ortakoy. The project consists of two buildings – one building is residential and the second is commercial. Nestled next to a hillside, the project will be an iconic step forward in the modernisation of the area.

The terrace roof of One & Ortakoy is handled as an element of the urban landscape. The design has one main concept: the lost area because of the construction had to be won on the roof. With a search for answers to "Can we swim on the roof?", "Can we run on the roof?", "Can we do yoga on the roof?", the main motivation of the project was to approach the roof as a party room, as a garden, as a new nature.

The roofs on both buildings are intended to be landscaped recreational playgrounds. These green roofs lessen the scale of the buildings, merging them with the existing topography. As well, the natural stone façade that wraps both buildings provides a soft skin blending with the natural hillside setting. The One & Ortakoy roof terrace plays a significant role in the perception of the building, as the roof is immediately visible upon approaching the building, in large part due to the project's proximity to the Bosphorous Bridge. Thus, the roof terrace is a dominant element, and is designed with this in mind. Conceived of as a green recreational level, the roof terrace includes a running track, swimming pool and extensive gardens. While the roof element serves a purpose to complete the building's form, the green roof also presents a smart approach for covering large reflective surfaces with vegetation, and therefore reduces carbon emissions for the building.

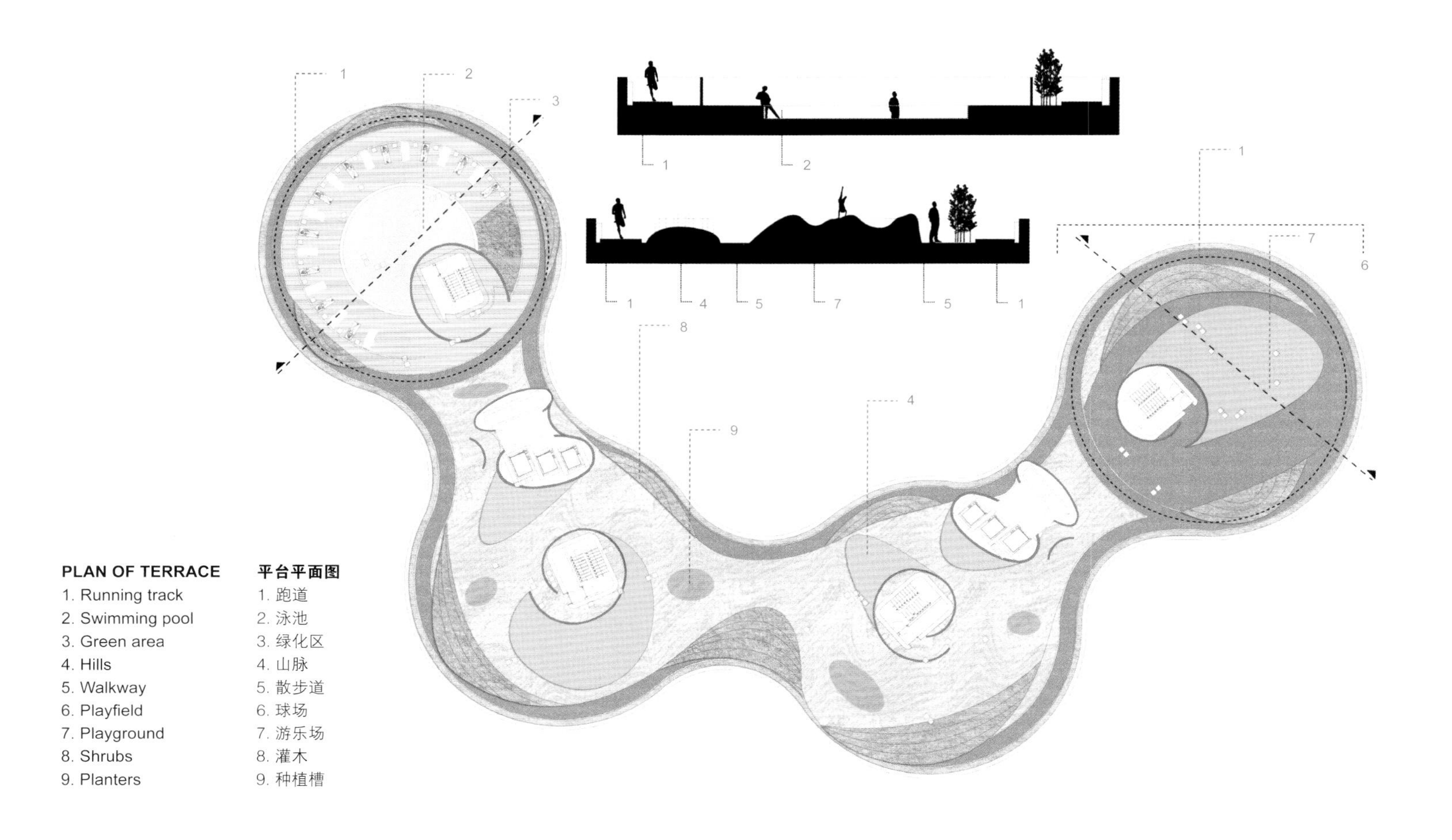

PLAN OF TERRACE　平台平面图

1. Running track　1. 跑道
2. Swimming pool　2. 泳池
3. Green area　3. 绿化区
4. Hills　4. 山脉
5. Walkway　5. 散步道
6. Playfield　6. 球场
7. Playground　7. 游乐场
8. Shrubs　8. 灌木
9. Planters　9. 种植槽

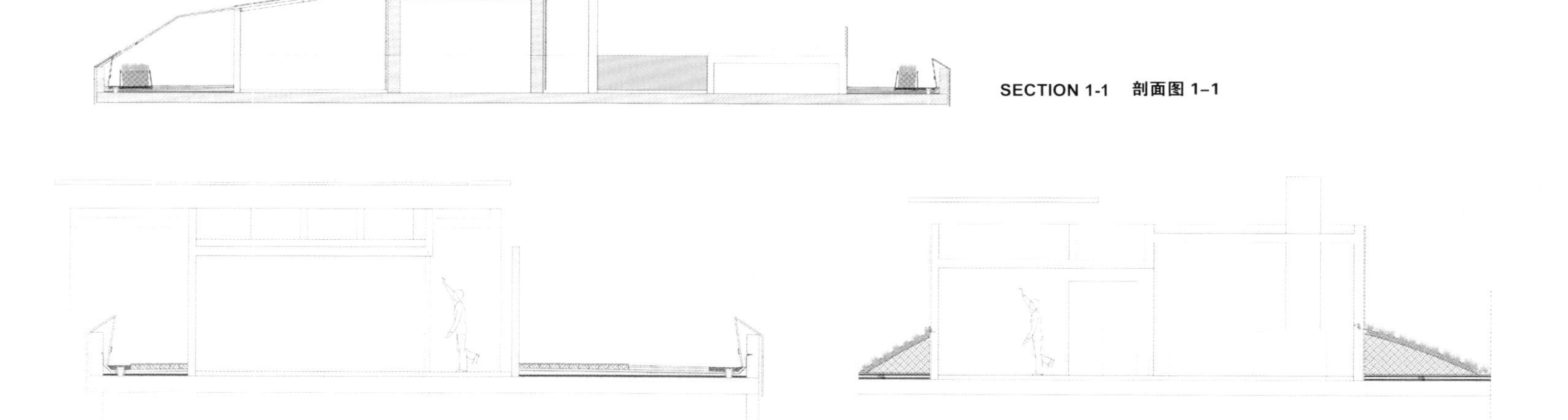

SECTION 1-1　剖面图 1–1

SECTION 2-2　剖面图 2–2

SECTION 3-3　剖面图 3–3

SECTION 4-4　剖面图 4–4

SECTION 5-5　剖面图 5–5

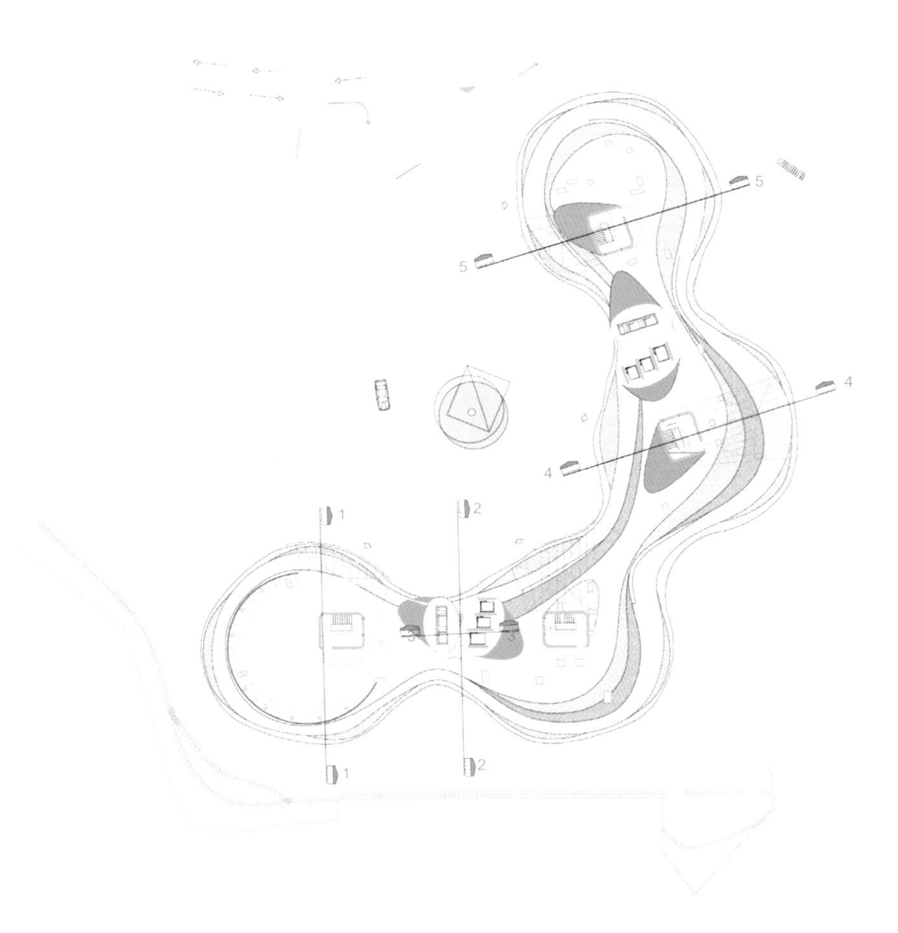

5
5
4
4
1
2
1
2

The landscape design for the roof makes use of a constructed artificial topography, inspired by the natural topography surrounding the building.

One & Ortakoy distinguishes itself from the neighbourhood, setting the project apart from other buildings in the neighbourhood. This is achieved, not only through the massing itself, but also by its relationship to the natural environment surrounding it. In essence, the building performs as a living organism, evolving from basement to roof terrace. Conceptually, the roof design serves to integrate the building with Ortakoy by dissolving into it, integrating duplex garden units in the basement and ground level with vertical gardens in the intermediate levels.

Therefore, it is the ambition and success of this project to be highly respectful of its natural surroundings, and to respond appropriately to the necessity for a sustainable design.

奥塔科伊综合楼位于伊斯坦布尔的奥塔科伊区，包含两栋建筑，一栋是住宅，一栋是商用，周围群山环绕。本案的开发是奥塔科伊区现代化建设的关键一步。

设计师将屋顶平台打造成城市景观的一部分。景观设计的理念是：由于建筑施工而失掉的绿化面积要在屋顶上找回来。“我们能在屋顶上游泳吗？”“我们能在屋顶上跑步吗？”“我们能在屋顶上练瑜伽吗？”设计师希望能给这些问题以肯定的回答，将屋顶平台打造成集花园和活动室于一体的优美环境，让大自然回归城市。

两栋建筑的屋顶都规划成绿化休闲区。屋顶的绿化从视觉上缩小了建筑体量，让建筑融入周围地貌中。另外，天然石材的外立面让建筑外观显得很硬，景观设计仿佛为其披上一件柔软的绿色外衣，让建筑与周围的山脉更加协调。屋顶平台的绿化对这两栋建筑的外观起到重要作用，因为当你逐渐走近这两栋建筑物的时候，最先看到的就是屋顶。因此，屋顶平台是一个决定性的元素。在设计过程中，设计师始终牢记这点。他们将屋顶平台定位在“绿色休闲空间”，规划了跑道、泳池和宽敞的园林空间。屋顶不仅是决定建筑外观的重要元素，而且还有一项重要功能，那就是用大面积的植被覆盖建筑表面，进而减少建筑物的碳排放。

屋顶平台上的地形地貌都是人工修建的，灵感来自建筑周围的自然地形。

奥塔科伊综合楼在该地区其他建筑物中脱颖而出，不仅由于这两栋建筑的造型与众不同，更重要的是因为景观设计使其融入了周围的自然环境中。这两栋建筑仿佛活的有机体，从地面自然地生长起来。从设计理念上说，屋顶的景观设计让建筑融入环境，进而成为环境的一部分，通过建筑外部的垂直绿化，将屋顶层与地面层自然地联系在一起。

因此，本案的设计充分尊重周围自然环境，设计的成功在于因地制宜，同时也满足了可持续设计的要求。

Prive by Sansiri

普里瓦高档公寓

Location:
Bangkok, Thailand
Landscape architect:
TROP: terrains + open space
Client:
Sansiri Venture Co., Ltd.
Photographer:
Charkhrit Chartarsa, Pok Kobkongsanti
Area:
760 sqm

项目地点：
泰国，曼谷
景观设计：
TROP景观事务所
客户：
盛诗里房地产公司
摄影师：
查克里特・查塔萨、波克・高贡桑蒂
面积：
760平方米

Project description:

Prive by Sansiri is an exclusive luxury condominium in the prime Bangkok location. The target group is a successful 40-plus people, so the design has to be neat and elegant.

TROP's scope includes the ground floor garden and the swimming pool on the roof.

For the ground floor, Sansiri asked TROP to create a wall to enclose and screen the lobby from the public. However, TROP found that the area is a bit small. So instead of building solid wall, which would make the area feel even smaller, they proposed a custom-designed sculpture wall as the alternative. The wall is a series of sculptural columns; each has some space between one another. As a result, the area has some ventilation and plays with natural light in a much more interesting way. At the base of the columns, TROP strategically placed a reflecting pond to make the visual even more beautiful.

For the pool, originally, the architects provided a small rectangular pool in the middle of the roof. Because there is a great view here, TROP suggested to create an L-shape pool, right at the edge of the building instead. With this design, one of the best view of Bangkok is offered for the resident. Then TROP played with the composition of the pool terrace. TROP divided the terrace into several portions. This way, a person would not see the whole garden at once. He has to walk around and discover some secret corners in the garden by himself. With a variety of space provided on the roof, everyone can use it without disturbing others.

ROOF PLAN 屋顶平面图

普里瓦高档公寓是泰国盛诗里房地产公司开发的楼盘，位于曼谷的黄金地段，目标住户是40岁左右的成功人士，所以设计上务求高雅奢华。

TROP景观事务所负责整个公寓的景观设计，包括一楼花园和楼顶泳池。

首先是一楼花园的设计。盛诗里公司要求TROP景观事务所设计隔断墙，将公寓前庭围起来，形成一道屏障。但是，设计师发现一楼空间较小。于是，他们没有采用常规的实心墙壁（这会让空间显得愈发狭小），而是专门打造了独特的雕塑般的镂空墙。镂空墙由一排柱子组成，彼此之间有一定间隔。这样一来，不仅有助于通风，而且在阳光照射下还形成妙趣横生的光影效

果。柱子下方设置倒影池，让整体环境显得愈加美丽。

其次是楼顶泳池的设计。建筑师原本在屋顶中央设计了一个长方形小泳池。由于屋顶视野很好，所以TROP景观事务所建议沿着建筑边缘打造“L”形泳池。这样，住户就能在泳池边欣赏曼谷的美丽风景了。在“L”形泳池的基础上，设计师重新规划了池边平台的空间，将平台拆分成几个部分。这样一来，就无法在一个点上看到整个屋顶的景致，必须漫步其间，慢慢欣赏每一个角落的美。由于空间的适当拆分，屋顶的使用更加方便，每个人可以偏安一隅，不会打扰到其他人。

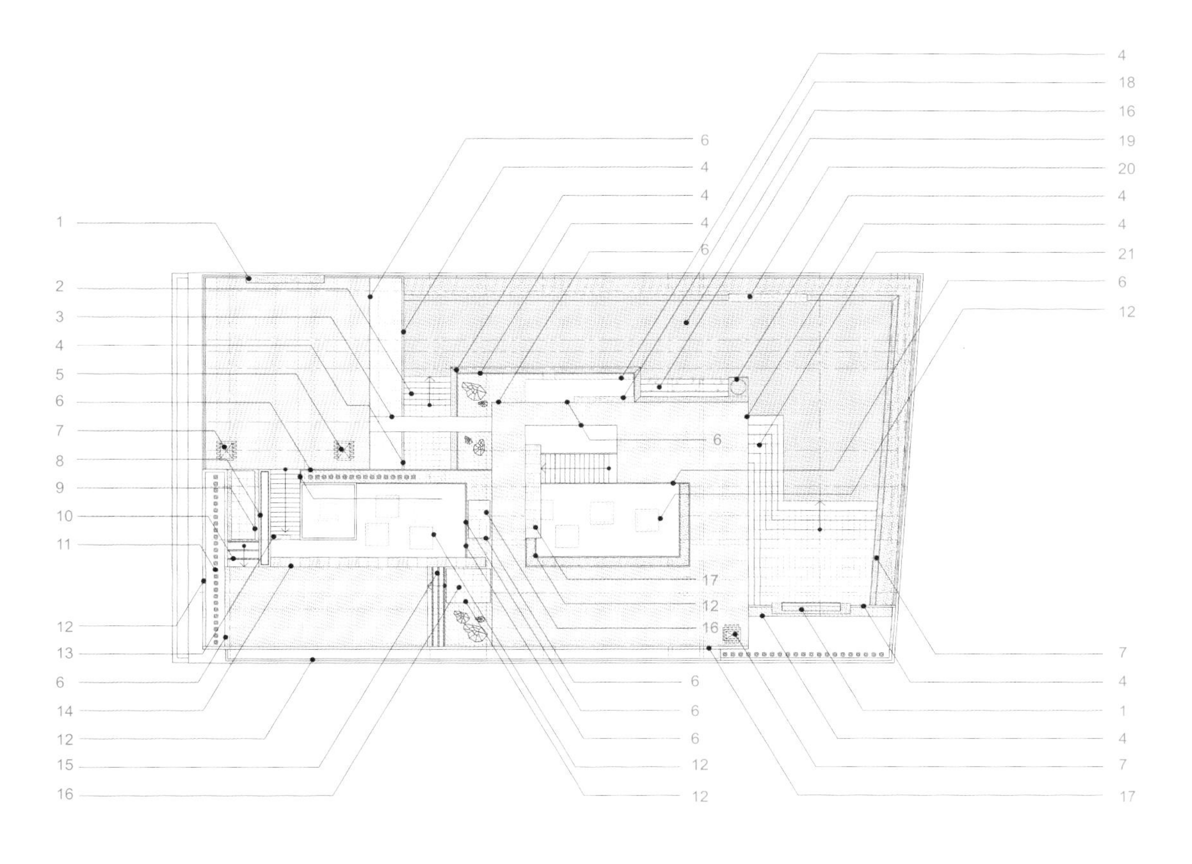

REFERENCE PLAN

1. Feature wall 6
2. Step 4
3. Stepping
4. Pond edge detail
5. Paving for manhold detail
6. Typical walkway / planter curb detail
7. Typical pond edge detail
8. Feature wall 1 / planter curb detail
9. Typical walkway
10. Step 2
11. Sculpture
12. Planter / planter curb detail
13. Feature wall 3
14. Feature wall 2
15. Step 1
16. Paving
17. Feature wall 4
18. Feature wall 5
19. Jacuzzi / pond edge detail
20. Feature wall 7
21. Step 3

参考平面图

1. 特色墙-6
2. 台阶-4
3. 踏步
4. 池塘边缘
5. 人工铺装
6. 标准走道/种植槽边缘
7. 标准池塘边缘
8. 特色墙-1/种植槽边缘
9. 标准走道
10. 台阶-2
11. 雕塑
12. 种植槽/种植槽边缘
13. 特色墙-3
14. 特色墙-2
15. 台阶-1
16. 铺装
17. 特色墙-4
18. 特色墙-5
19. 极可意按摩浴缸/泳池边缘
20. 特色墙-7
21. 台阶-3

MATERIAL PLAN

1. Concrete white colour paint toa super shield 8260
2. Laminated glass rail by architect
3. Loose black pebble stone size Φ3-5cm
4. Black China granite stone thick 2.5, 5cm
5. Black granite basalt stone thick 2.5, 5cm
6. TSM 1
7. TSM 2
8. Loose black slate stone, natural shape size 10, 15cm. PROX
9. Stainless steel, thick 1cm
10. Black volcanic stone
11. Travertine stone
12. Teak wood floor plank
    size: 0.70x0.10x0.05m thk.
    size: 0.10x0.10x0.05m thk.

材料平面图

1. 白色喷漆混凝土板（8260）
2. 夹层玻璃栏杆（由建筑师设计）
3. 黑色鹅卵石（直径3厘米～5厘米）
4. 黑色中国花岗岩（厚度：2.5厘米或5厘米）
5. 黑色玄武岩（厚度：2.5厘米或5厘米）
6. TSM-1
7. TSM-2
8. 黑色石板（天然形状，规格为约10厘米或15厘米）
9. 不锈钢（厚度：1厘米）
10. 黑色火山岩
11. 石灰华
12. 柚木地板（规格1：0.70米x 0.10米x 0.05米；
    规格2：0.10米x 0.10米x 0.05米）

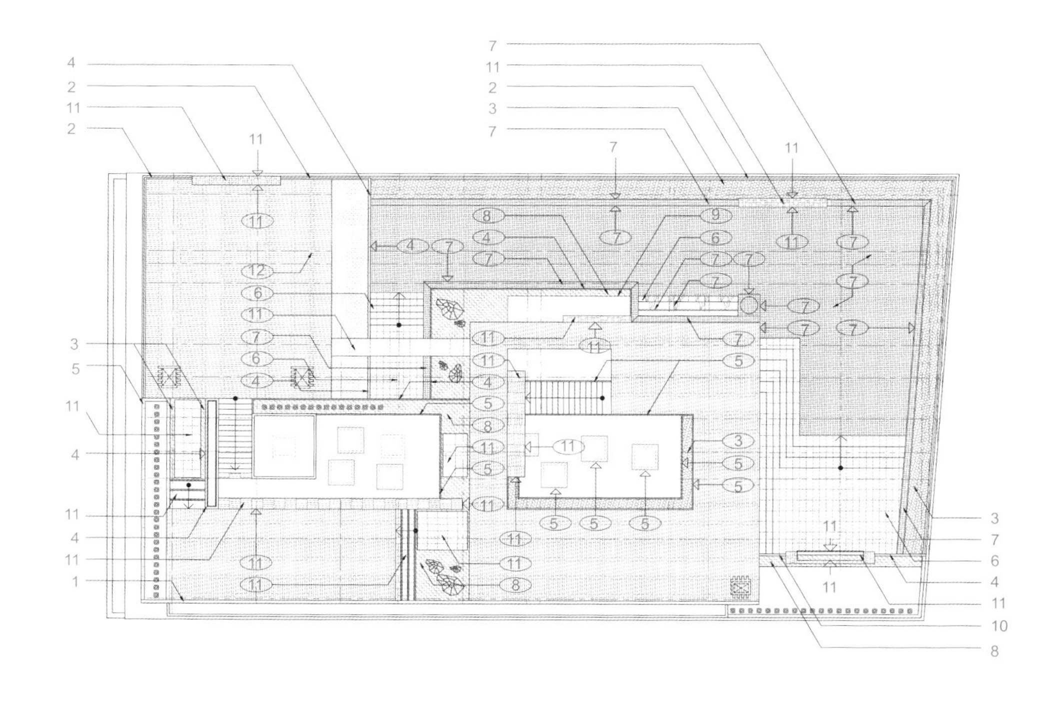

Roof Garden Praderas

普拉德拉斯屋顶花园

Completion date:
2011
Location:
Mexico City, Mexico
Architect:
Arquitectura de Paisaje
Photographer:
Kees Van Rooij
Area:
2,800 sqm

竣工时间：
2011年
项目地点：
墨西哥，墨西哥城
建筑设计：
"建筑景观"事务所
摄影师：
凯斯·凡罗伊
面积：
2,800平方米

Project description:

The design is inspired by the reflection of the planetarium on the roof top of a corporate building. It allows privileged views of the Pedregal de San Angel and of the national park-reserve of Pedregal and the unique opportunity to integrate construction with lead standards.

The fifth façade is thought as an additional programme to coordinate architecture, structure and landscaping.

The final slab is built adapting to slopes optimising pluvial drainage system. Imperceptible details such as the arrival of the lift directly to the roof garden or the rhythm of the structure reflected on the location of the trees, creating scale and solving structure problems.

The roof garden generates microclimates through the pergola, as well as a permeable character that becomes the joining concept on the design.

Even though the finish of the project is porous and of stone materials, it has been prepared for the comfort of the user; you are even able to walk on high heels.

Since there are no areas of grass and bushes, the maintenance is very low cost and the plant species are mostly endemic to the region.

The roof garden becomes a multi-functional amenity that can be used for a casual relaxing walk or an event for a thousand people, providing a new experience for the corporate world.

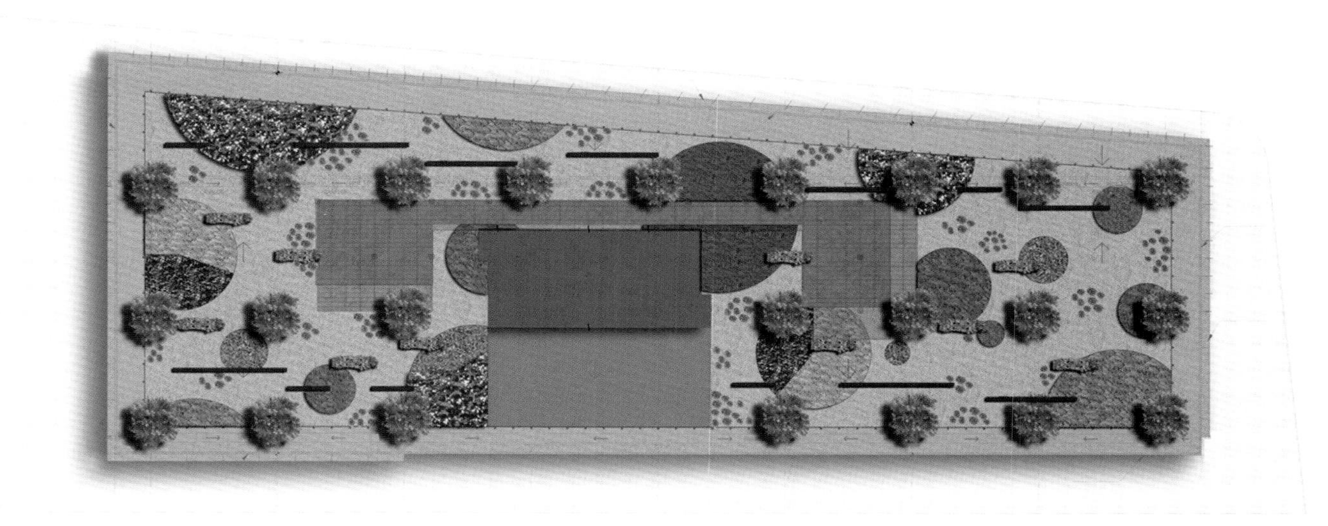

ROOF FLOOR PLANTING 屋顶植被示意图

本案的设计灵感来自于星象仪在这家公司的办公楼楼顶上的反射图像。从这里可以眺望古老的佩德雷加尔住宅区和佩德雷加尔国家级自然保护区的风景。屋顶的景观设计与建筑结构完美地融为一体。

除了屋顶之外，5楼的立面也进行了绿化，也体现了景观和建筑结构的有机结合。

屋顶的板材成一定角度铺设，形成斜坡，以便解决雨季的排水问题。设计师的细心体现在许多细节中。比如，从电梯出来，直接就到达屋顶花园；再比如，树木的栽种位置和建筑结构的韵律相互呼应，营造出适当的尺度和比例，解决了景观空间的结构问题。

屋顶花园上的棚架有助于建立微气候。棚架也是建筑和景观之间的一个衔接元素。

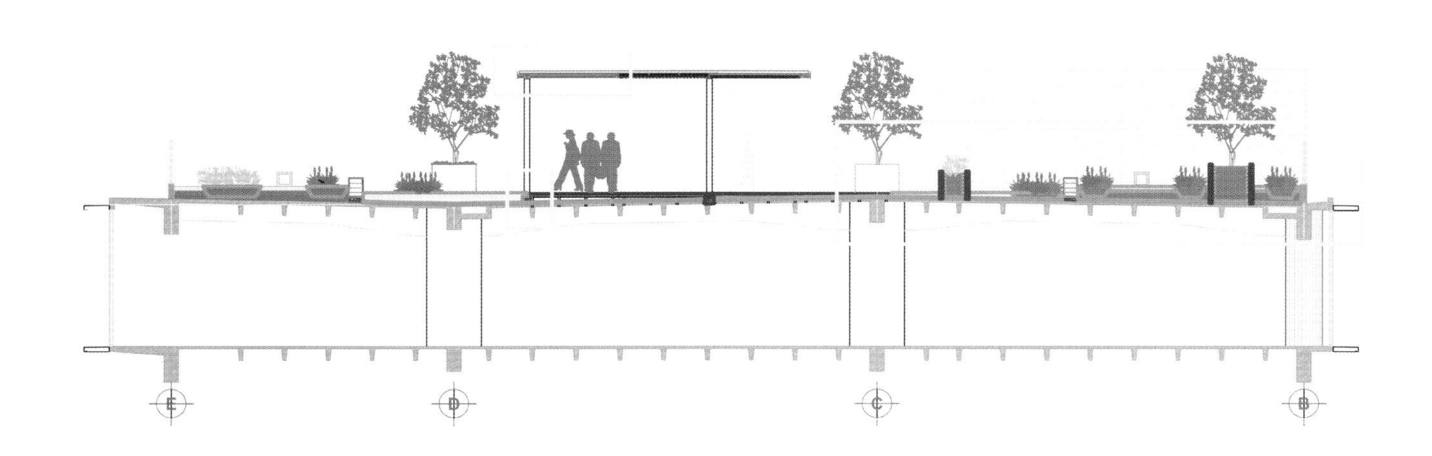

SECTION 剖面图

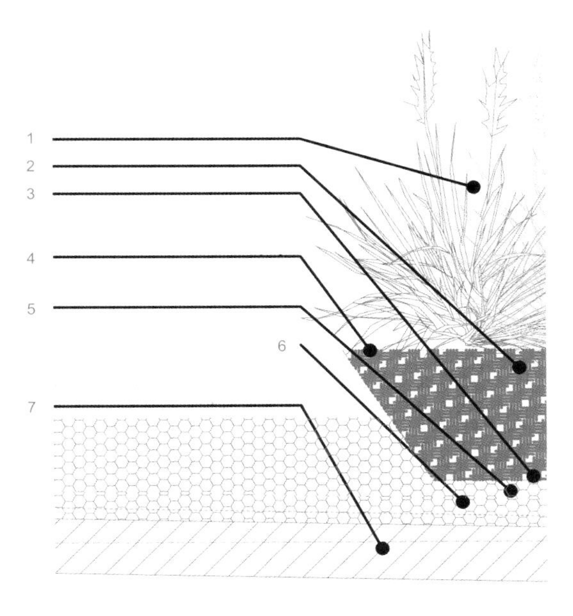

DETAIL 1

1. Vegetation/ground covers
2. Soil
3. Geotextile
4. Tepjal
5. Tezontle
6. Gravel of 2"
7. Slab

细部图-1

1. 植被/地表植物
2. 土壤
3. 土工织物
4. Tepjal火山岩
5. Tezontle火山岩
6. 砂砾（2英寸，约5厘米）
7. 板材

DETAIL 2

1. Citric tree
2. Concrete
3. Soil/geotextile
4. Gravel drain
5. Slab
6. Recinto bench
7. Wood bench
8. Tepjal
9. Tezontle
10. Gravel of 2"

细部图-2

1. 柠檬树
2. 混凝土
3. 土壤/土工织物
4. 砂砾排水渠
5. 板材
6. 围栏长椅
7. 木质长椅
8. Tepjal火山岩
9. Tezontle火山岩
10. 砂砾（2英寸，约5厘米）

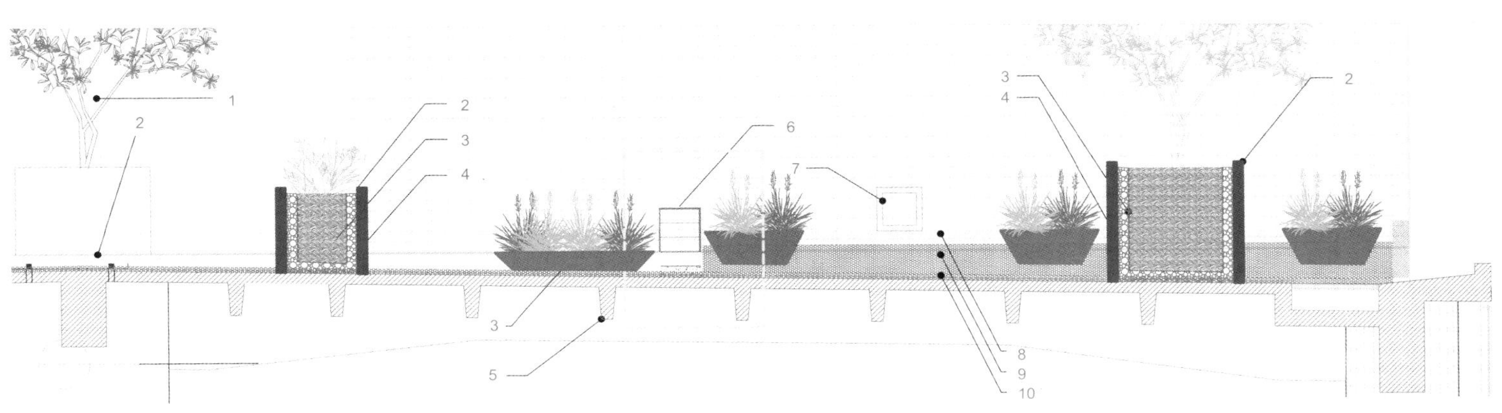

尽管本案中使用的材料是有渗水孔的，主要是石材，但设计师还是考虑到了使用舒适性；即使穿着高跟鞋也能在屋顶上自如行走。

屋顶上没有草坪和灌木，所以维护成本很低，植物种类大部分都是本地植物。

这座屋顶花园是一个多功能休闲空间，既可以在这里轻松地散步，也可以进行集会活动（能容纳上千人），为这家公司的办公生活带来全新体验。

Sydney City Rooftop

悉尼屋顶花园

Location:
Sydney, Australia
Landscape architect:
Secret Gardens of Sydney
Photographer:
Jason Busch

项目地点：
澳大利亚，悉尼
景观设计：
"悉尼秘密花园"设计公司
摄影师：
贾森·布什

Project description:

This apartment is actually the floor below the penthouse that Secret Gardens had completed the year prior.

The design focused on renovating an existing garden that was currently obstructing the views from this city apartment. The style had to be consistent with the architecture of the building and the client's preference was a garden that was more formal in its approach. This was created with clean lines of Buxus hedging and topiary cones. The introduction of Iris and Lavender, softened the formal aspects and created a whimsical effect.

The existing tiles on the balconies, and none of the external architecture of the building could be modified so all materials and plants selected need to be complimentary to these materials.

The result is a beautiful garden located 25 floors in the sky! Clean lines and a simple design, have created a spot to place a chair, sit in the middle of the city and enjoy the view. The low levels of hedging ensure the clients can now see the amazing view of Sydney harbour and the Opera House.

With stringent building codes on what could and could not be done, the designers worked with the building management to ensure that the design and construction could meet the client's brief. Everything on site had to be brought up in a lift so the design had to cater for this.

The turf is artificial and therefore does not require water. The plant beds were less than 300mm so drainage cell was used in the planters to create a cavity for water flow. The garden beds are run on a drop irrigation system. The garden has now been in place for about four years and is thriving with very little maintenance. The designers maintain the garden every eight weeks.

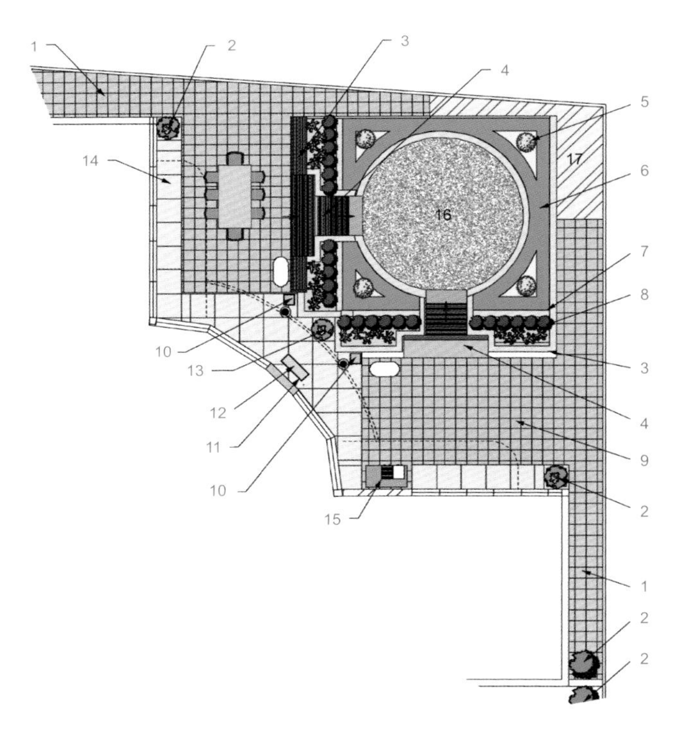

MASTERPLAN

1. Existing tiles retained
2. 700x700mm metal planter
3. Custom-built classic style Tallowood bench seat over edge of existing raised garden bed
4. 86x19mm Tallowood deck to extend over first tread of existing steps. Selected oil finish
5. Buxus topiary balls
6. Retain existing Buxus hedge
7. Existing paving strip between gardens to be removed
8. Remove existing Murraya hedge
9. Existing tiles to be lifted and re-used after waterproofing slab
10. 400x400mm metal planter
11. Relocate services from wall
12. 1100x390x700mm GRC planter in "Charcoal" to support Star Jasmine up wall and over pergola
13. 800x400mm contemporary bowl in "Charcoal"
14. 900x900mm granite tile to match existing on "Versipave" system
15. Electrolux style BBQ
16. Existing lawn retained
17. Void

总平面图

1. 保留了原有的地砖
2. 金属种植槽（规格：700毫米×700毫米）
3. 特制的古典风格脂木长椅，安装在原有的花池边缘上
4. 脂木平台，作为原有台阶的第一级踏步的延伸；脂木浸油处理
5. 修剪成球状的黄杨
6. 保留了原有的黄杨树篱
7. 拆除了绿化区之间原有的带状铺装
8. 拆除了原有的九里香树篱
9. 原有的地砖进行再利用；下面铺一层防水板
10. 金属种植槽（规格：700毫米×700毫米）
11. 围墙材料再利用
12. 玻璃纤维增强水泥（GRC）种植槽（规格：1100毫米×390毫米×700毫米，里面种植的络石会攀爬到墙壁和棚架上）
13. 现代化种植槽（规格：800毫米×400毫米）
14. 花岗石砖（规格：900毫米×900毫米），与原有的铺装材料风格一致
15. 欧式风格烤肉设备
16. 保留了原有的草坪
17. 中空

POT SCHEDULE 种植槽		
QTY 数量	PLANTER 样式规格	FINISH 表面材料
2x	400x400x600mm custom metal "Versailles" planters 特制金属种植槽；400毫米×400毫米×600毫米	Treasury Bronze 铜
4x	700x700x700mm custom metal "Versailles" planters 特制金属种植槽；700毫米×700毫米×700毫米	Treasury Bronze 铜
1x	800x400mm GL contemporary bowl 现代风格种植槽；800毫米×400毫米	Charcoal 炭
1x	GRC 1100x390x700mm trough planter 玻璃纤维增强水泥长条种植槽；1100毫米×390毫米×700毫米	Charcoal 炭

Plants used included: Buxus microphylla - Japanese Box; Dietes robinsoniana - Wedding Lilly; Raphiolepsis indica “Snow maiden” - Indian Hawthorne; Lavandula stoechas “Avon View” - Italian Lavender; Juniperus chinesis - Spartan Juniper; Trachelospermum asiaticum - Star Jasmine; Gardenia augustifolia “florida” - Gardenia.

本案是一间公寓的屋顶花园设计。“悉尼秘密花园”设计公司前一年刚刚打造了这户楼上的顶层公寓的屋顶花园。

本案的设计聚焦原有屋顶花园的改造——原来的屋顶主要问题是视野受阻。屋顶的设计风格要与这栋公寓楼的建筑风格一致。房主喜好空间条理分明的布局。为此，设计师采用黄杨树篱和修剪成锥形造型的植被，呈现出简洁的线条。薰衣草的使用柔化了布局的僵硬棱角，营造出曼妙的整体氛围。

阳台上原有的瓷砖等所有外部建筑材料都不能改动，所以选用的所有材料和植物要与原有材质形成良好的互补效果。

在上述条件的制约下，设计师打造了一座25层高的美丽的空中花园。清晰的线条，简洁的设计，放上一张躺椅，坐在悉尼正中心欣赏全城美景，这里是再好不过的地方了。树篱比较低矮，不会阻挡欣赏风景的视线，在这里可以把悉尼港和悉尼歌剧院的全景尽收眼底。

因为当地有严格的建筑条例，详细规定了什么可以做，什么不可以做，所以设计师与当地建筑管理部门紧密合作，确保设计与施工能够在不违反相关规定的情况下满足业主的要求。施工中用到的所有材料都要用一部电梯运送到屋顶上，这一点也是设计师必须考虑到的。

屋顶上采用人造草坪，无需灌溉。花池宽度不大于300毫米，种植槽内采用排水槽，确保水可以在其中流动。花池采用滴注式灌溉系统。到目前为止，这个屋顶花园已经竣工4年了，目前植物长势良好，无需过多的维护。设计师每8周对屋顶花园进行一次维护。

所用植物包括：小叶黄杨、野鸢尾（婚礼百合）、石斑木、意大利薰衣草、柏树、亚洲络石、栀子花等。

Botanical Name 植物学名	Common Name 俗称	Scheduled Size 预计尺寸	Quantity 数量	Comments 备注
Buxus japonica 'Ball' 球状日本黄杨	Box "盒子"	400mm	2	
Buxus japonica 'Cone' 锥状日本黄杨	Box "盒子"	500mm	4	
Buxus japonica 'Cube' 方形日本黄杨	Box "盒子"	400mm	8	
Juniperus viginiana 'Spartan' "斯巴达"桧	Spartan Juniper 斯巴达杜松	500mm	4	
Moraea robinsoniana 肖鸢尾	Native Iris 北美鸢尾	140mm	32	
Neomarica caerulea 巴西鸢尾	Walking Iris "散步鸢尾"	140mm	2	
Raphiolepis 'Snow Maiden' "雪姑娘"石斑木	Dwarf Hawthorn 矮山楂树	400mm	20	
Trachelospermum asiaticum 亚洲络石	Dwarf Star Jasmine 矮星络石	140mm	26	
Trachelospermum jasminoides 络石	Star Jasmine 白花藤	140mm	2	grow on wire work 在金属丝网栏上攀援生长

Note: The establishment period is a critical time in the development of a new landscape. New plantings need to be watered, fertilised, and pruned to shape until they have become hardy enough to survive without help from people. Some plants will always need some extra care to survive.

注意：植物的生长成熟期对一个新的景观来说是至关重要的时期。新栽种的植物需要浇水、施肥、修剪，直到长到足够成熟，不需人力的帮助即可存活。有些植物成熟后仍需要进行特别的人工照料。

PLANTING PLAN

1. Juniperus viginiana 'Spartan'
2. Trachelospermum asiaticum
3. Buxus japonica 'Ball'
4. Clients existing Gardenia in bowl
5. Neomarica caerulea
6. Trachelospermum jasminoides grow on wire work
7. Raphiolepis 'Snow Maiden'
8. Moraea robinsoniana
9. Buxus japonica 'Cube'
10. Buxus topiary cones in centre areas of hedge. Allow existing hedge to grow in to fill gaps around new 'cones'
11. Maintain Raphiolepis slightly higher than Buxus
12. Moraea robinsoniana

植物平面图

1. "斯巴达"桧
2. 亚洲络石
3. 球状日本黄杨
4. 客户原有的栀子花
5. 巴西鸢尾
6. 爬满金属丝网栏的络石
7. "雪姑娘"石斑木
8. 肖鸢尾
9. 方形日本黄杨
10. 树篱中央的锥状日本黄杨（原有的树篱继续生长，逐渐填补这些黄杨中间的空隙）
11. 石斑木（保持修剪在比黄杨略高的高度）
12. 肖鸢尾

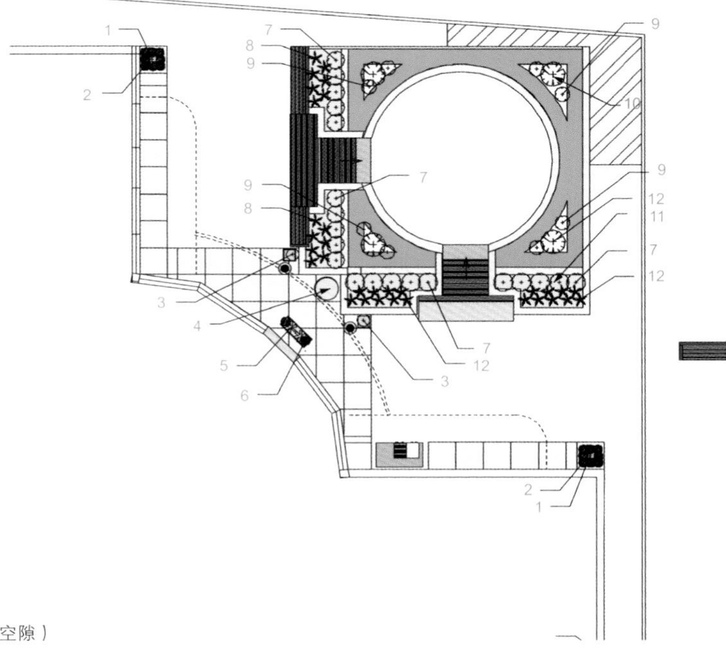

The Pool @ Pyne

派恩公寓空中泳池

Completion date:
2013
Location:
Bangkok, Thailand
Architect:
Palmer & Turner (Thailand) Co., Ltd.
Landscape architect:
TROP: terrains + open space
Photographer:
Wison Tungthunya
Area:
370 sqm

竣工时间：
2013年
项目地点：
泰国，曼谷
建筑设计：
泰国P&T建筑设计集团
景观设计：
TROP设计工作室
摄影师：
威森·唐森亚
面积：
370平方米

Project description:

Bangkok has changed. So have its people.

In the past, we may prefer to live in small houses outside the city areas, and commute in and out the city daily. Not anymore. To fit the present time's fast lifestyle, new generation keeps moving in many condominiums inside the developed areas instead.

Horizontal Living is out. Vertical one is the thing to do.

As a result, Thai developers are competing hard for the perfect plots of land in town. No, they do not care much about how big the plot is, or how great the view it would get. As long as it is right next to the BTS (Bangkok's Sky Train) station, it is perfect. In 2010, TROP got a commission to design the Pool of Pyne by Sansiri, a high-end condominium in Bangkok. Its site is ideal. Located right in the middle of busy urban district, just 5 mins walk from the city's biggest shopping malls, the plot is about the right size, 2,900sqm. To make it even better, it also has a BTS station right in front of the property.

Architecture-wise, most condominiums in Bangkok are quite similar. The residential tower is built on top of parking structure. Normally the parking part has a bigger floor plan than the tower, leaving the left-over area as its swimming pool. The Pool @ Pyne by Sansiri is no different. It is designated to be on the 8th floor, which is also the roof of the parking structure. The area is a rectangular shape terrace, around 370sqm.

Having the train station right in front really helps selling residential units (sold out in one day). However, space-wise, the station is a nightmare for designers. It is designed as a

huge structure, about 100 metres long, 3-4 storeys high. Basically, it is like placing a huge building right in front of your door steps. Together with other surrounding old buildings, this project is trapped among concrete boxes by all four sides.

In order to get rid of that boxy feeling space, TROP's first move is to create a "loosed" floor plan. Instead of a typical rectangular pool deck, they proposed a series of smaller terraces integrated with the swimming pool. Perpendicular lines were avoided, replaced by angled ones with round corners. A series of "green" planters were also inserted here and there, combining all three elements, water, terraces and plantings seamlessly.

Again, most pools in Bangkok share the same name. They are called "Sky Pool", because of their location on top of the roof. The first couple ones sounded very exciting, but, after a while, it got boring. The design task was not only to design a pretty swimming pool, but TROP also wanted to create a unique landscape feature that can identify the character of the residents.

To make the pool different than others, the "Skeleton", a light clad structure, was proposed to "frame" the swimming pool three-dimensionally. Before, the so-called sky pool is just a flat piece of water on top of the building. Sure, swimmers can enjoy a great prospect view outside, but, looking back to the building, nobody recognise the presence of that pool from below. With the "Skeleton", the pool was fully integrated into the architecture. Now the BTS passengers can look up and see the special space inside the frame. At night, the "Skeleton" glows, giving the architecture some "lightness" it needs badly in the crowded surrounding.

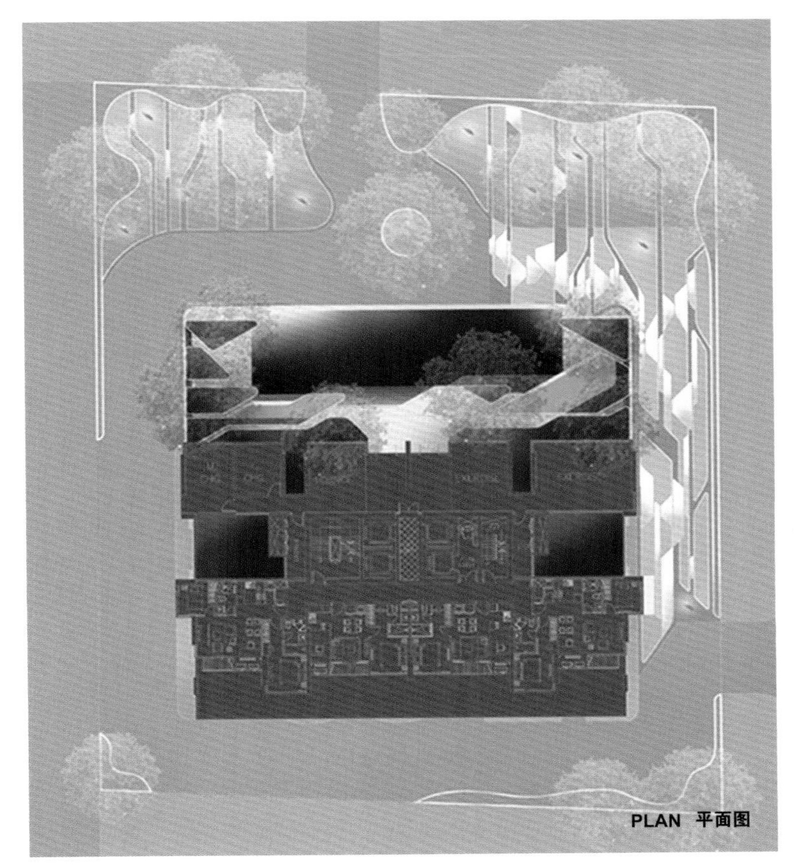

PLAN 平面图

曼谷变了。曼谷人也变了。

过去，我们可能更喜欢住在市郊的小别墅里，每天上下班通勤来往于市区与城郊之间。现在却不是了。为了适应当今社会更快的生活节奏，年轻一代选择住在市区的公寓里。

水平式的居住模式已经过时了。现在是垂直住宅当道。

于是，泰国的开发商正在为在市区的好地段而竞争。他们不在乎地块大小，也不管视野如何。只要离曼谷高架电车（BTS）车站近，就是好地段。2010年，TROP设计工作室接到设计任务，负责打造由盛诗里开发的一栋高档公寓的空中泳池。派恩公寓位于曼谷繁华的商业区，地理位置优越。从这里走到曼谷最大的购物中心只需5分钟。这一地块的大小也正合适，占地2,900平方米。尤其锦上添花的是，公寓对面就是高架电车站。

曼谷大部分的公寓楼从建筑设计的角度看都很相似。住宅楼的下层一般都是停车场。一般来说，停车场的占地面积会比上方的住宅楼大，形成裙楼的格局。裙楼顶层便可设计成泳池。派恩公寓的空中泳池也不例外。这个泳池地处8层高的位置，也就是停车场的屋顶。整个屋顶是个长方形的平台，面积约为370平方米。

公寓对面是高架电车站，这确实有利于卖房（一天之内就销售一空）。但是，对设计师来说，从空间设计的角度来看，车站却是个噩梦。这个车站是一栋大体量的建筑，长约100米，3～4层高。这就相当于在你家门口建起一座高楼。再加上周围的其他老旧建筑，这栋公寓被大体量的混凝土建筑物四面包围。

为了摆脱这种"遭围困"的局面，设计师采取的第一步就是拆分空间。设计师没有采用常规的长方形宽敞平台，而是将空间拆分成一系列小平台，与泳池融为一体。设计师避免出现垂直的线条，而是采用圆角和斜线。此外，平台各处还设置了一系列花坛，将水、平台和植物这三大元素合而为一。

曼谷大部分这类泳池都有一个共同的名称——"空中泳池"，因为地处屋顶之上。最初，刚有一两个这种泳池的时候，这个名字显得很吸引人。但是，随着数量越来越多，就变得千篇一律，令人厌烦了。设计师希望跳脱出这个常规模式，不仅要设计出漂亮的泳池，而且要打造独特的景观，让派恩公寓的住户能够因此而独树一帜。

为了让这个泳池与众不同，设计师采用了一个三维立体结构作为整个泳池的框架。从前，所谓的"空中泳池"只不过是屋顶上的平面水池。虽然人们能够眺望风景，但是，回过头看看建筑本身，从下面根本无法辨别哪个泳池是哪栋楼的。有了这个三维框架结构，情况就不同了，泳池与建筑完全融为一体。现在，高架电车的乘客可以抬头仰望，能看到框架结构内的独特空间设计。夜晚，框架结构在照明效果的烘托下发出光芒，让整栋建筑显得更加轻盈——这在拥挤的建筑环境中显得尤为重要。

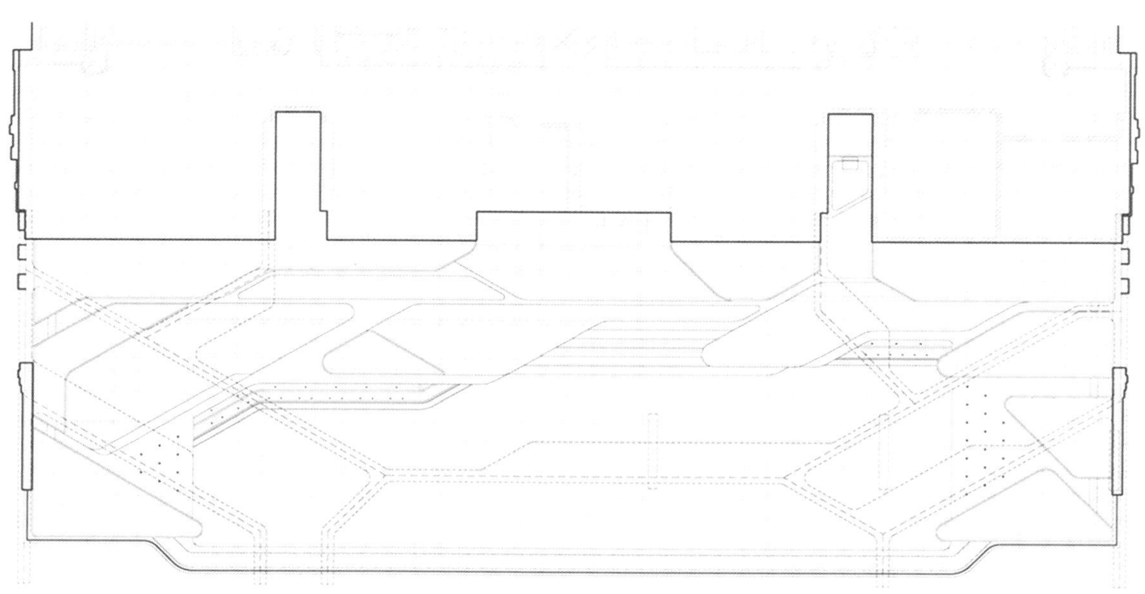

PLAN 平面图

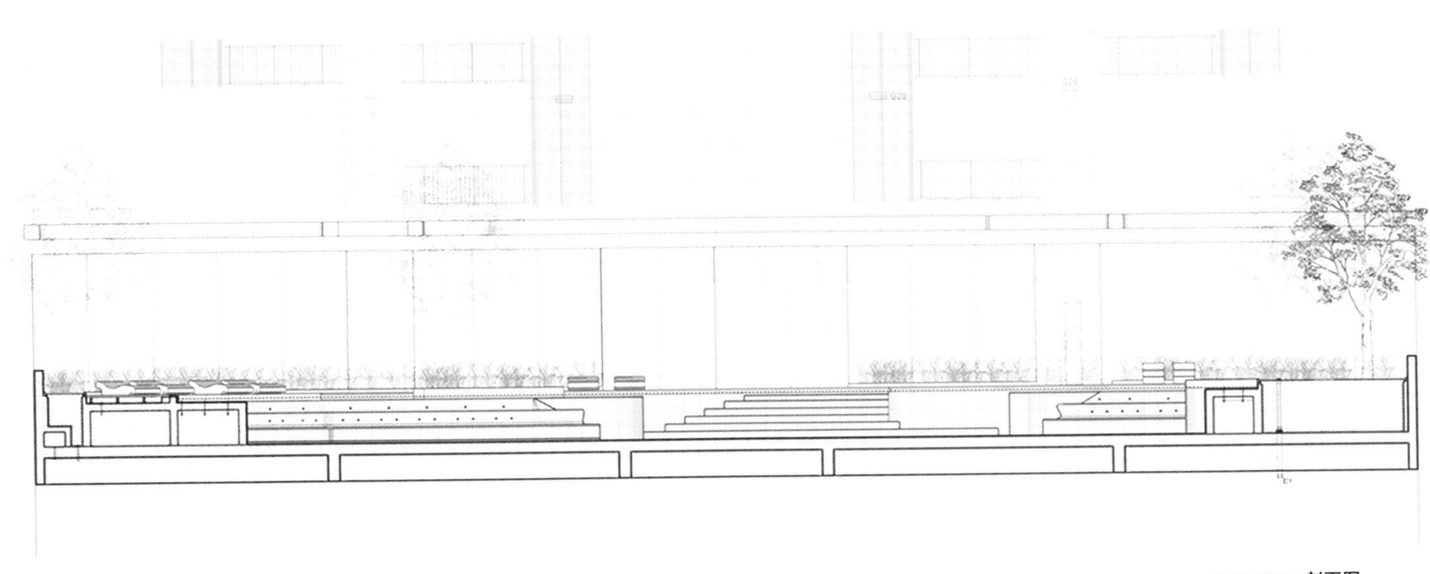

SECTION 剖面图

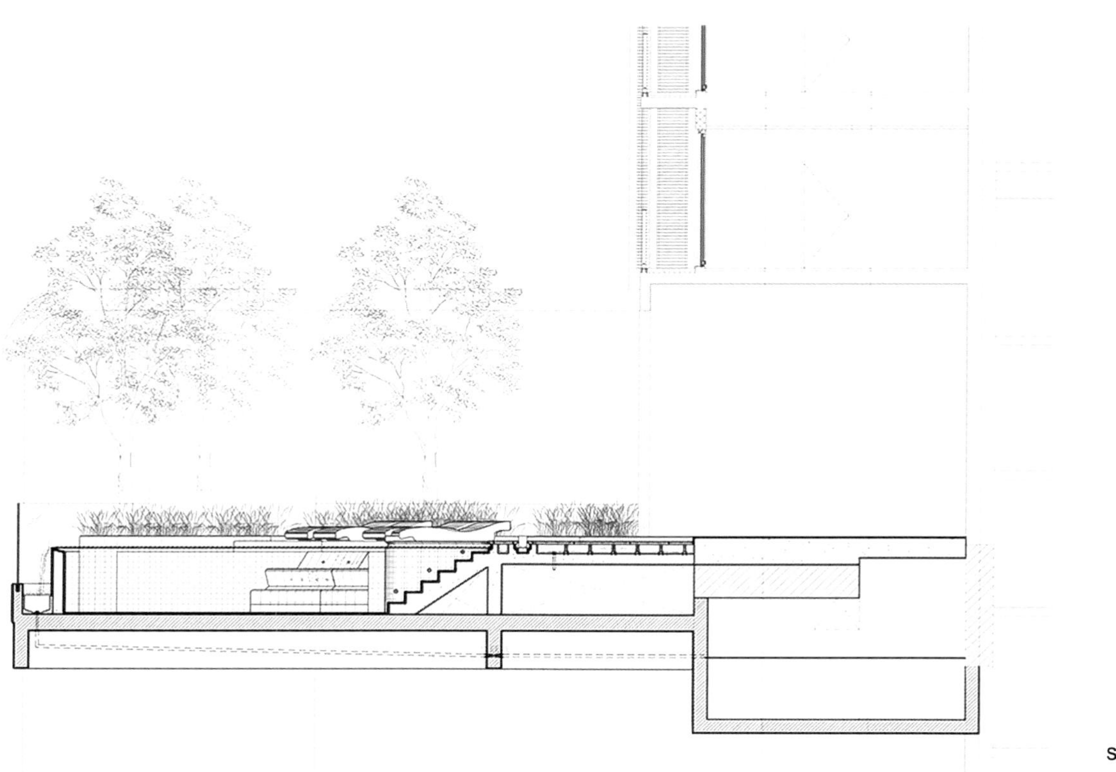

SECTION 剖面图

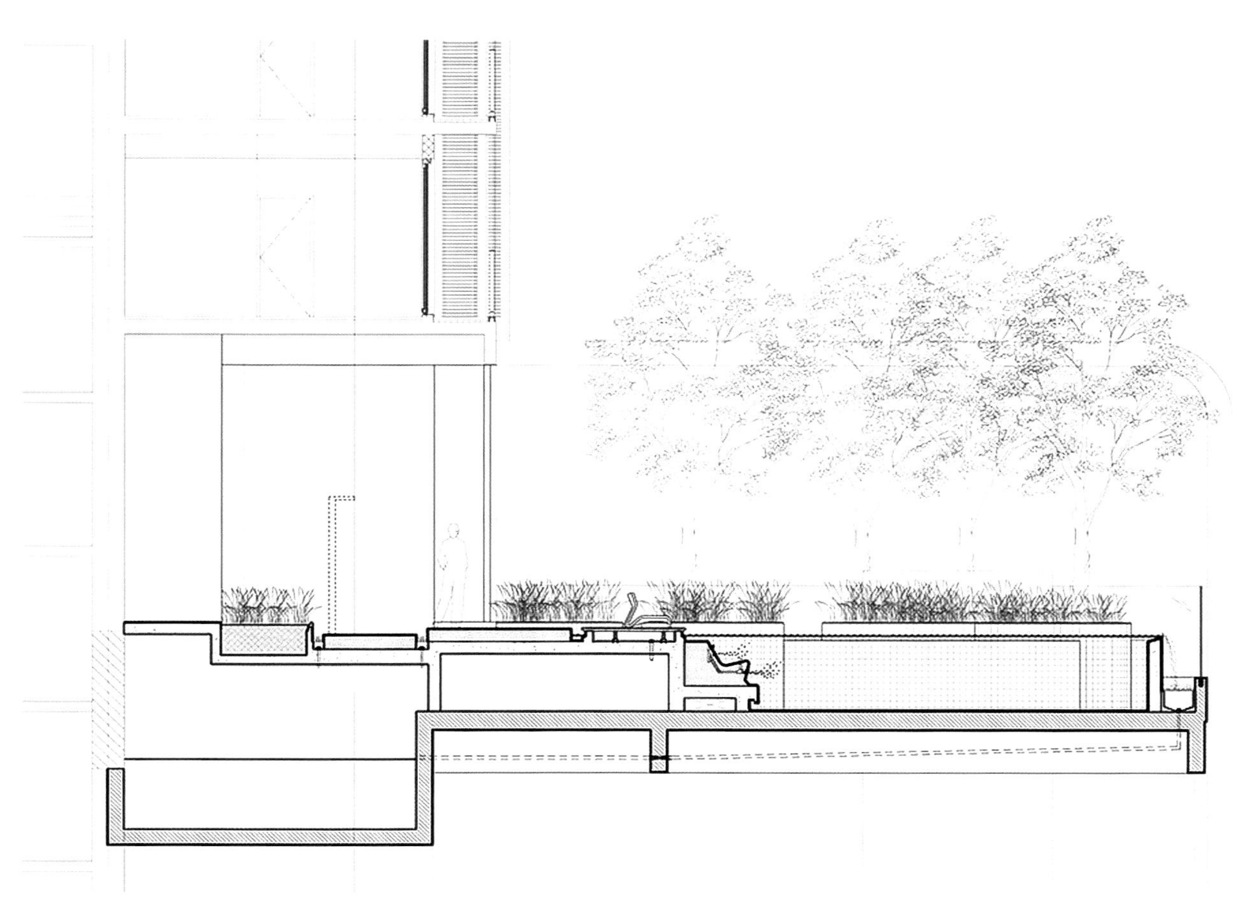

SECTION 剖面图

Tokyu Plaza Omotesando Harajuku

原宿表参道大街空中广场

Completion date:
2012
Location:
Tokyo, Japan
Architect:
Hiroshi Nakamura & NAP
Structure design:
Takenaka Corporation
Photographer:
Hiroshi Nakamura & NAP
Area:
1,770 sqm

竣工时间：
2012年
项目地点：
日本，东京
建筑设计：
中村拓志建筑事务所
结构设计：
竹中工程公司
摄影师：
中村拓志建筑事务所
面积：
1,770平方米

Project description:

Omotesando, an avenue lined with flagship shops of international fashion brands, is one of Japan's foremost fashion streets. As the frontal ("Omote") approach to the wooded compound of Meiji Shrine, the avenue is lined on both sides with zelkova trees. The branches of the trees form a canopy over the avenue and provide an experience of shopping under sunlight filtering through trees. An environment that blends architectural spaces with spaces under trees – such is the special character of Omotesando, and Hiroshi Nakamura & NAP therefore proposed a style of environment giving emphasis to this special character.

In concrete terms, Hiroshi Nakamura & NAP sought to give the space of the avenue below the trees a phased connection with the upper portion of the building, by floating in the sky a volume that merges the building with the zelkova trees. This not only allows birds and butterflies to easily ascend to the roof-top forest; a roof garden there also gives people an incentive to visit the building. Commercially, a shower-like effect of customers spilling down from the top floor to the floors below is produced.

Hiroshi Nakamura & NAP first laid out terraces and located trees on the roof and in peripheral spaces – areas of the building receiving maximum sunlight and, moreover, enjoying unlimited views of the intersection. They then created a volume three or more storeys in height. As a result, the spreading branches of trees displaced portions of the building's interior, and the terrace had to be elevated to the height of the trees' root clumps. They nevertheless chose to embrace these restrictive conditions and use them positively to set up a feeling of intimacy between the trees and visitors. By establishing polygonal steps on the terrace, similar in

appearance to the polyporaceae (bracket fungi) that grow at the base of trees, they could fill in the height difference created by the root clumps and produce a terrace of mortar shape. The polygonal steps have a multiplicity of meanings, for they serve not only as steps but also as chairs and tables. Through them, users enter a physical dialogue with the building.

What particularly charmed the designers about the mortar-shape terrace is how visitors naturally gravitate to the lower levels of the terrace, seeking a centre. There, the visitors seat themselves in the countless nooks around the terrace and share the scenery before them. In these times, when Internet shopping is on the rise, it is important for commercial spaces to maximise the value they possess that cannot be replicated online. The relaxed sense of unity one feels in this comfortable environment abounding with distinctive places and pleasant physical experiences is precisely the quality demanded of commercial facilities in the future.

A Kaleidoscope-like Entrance Tube

Memory of the site's past has been maintained in the entrance tube leading visitors to upper floors. The tube is a great mirrored space, its mirrors recalling the mirror wall cladding of the Central Apartments, which once occupied the site (a building that came to symbolise Sixties and Seventies youth culture). To see fashionably costumed people reflected repetitively in the mirrors, like colourful objects in a kaleidoscope, is dazzling in effect. A special sense of excitement evocative of the fashion world is produced, and the ride on the escalator becomes a rich experience. Attracted by the mirrored reflections of people passing through the intersection, one peers into the tube. On riding up the escalator, a great atrium appears with tree-filtered light spilling down from a skylight in the ceiling. Thus begins a shopping experience true in character to Omotesando.

Employing Light to Produce a Façade of Varying Expression

The external wall cladding is a composition of polygons that

表参道大街是东京原宿区的一条繁华大道，道边国际时尚品牌旗舰店林立，是日本最知名的时尚一条街。东京著名的明治神宫就面向表参道大街。明治神宫采用木质结构，而表参道大街两边种植了成行的榉树，与之呼应。树木的枝干在街道上方形成遮篷，阳光透过枝叶洒下来，带来别样的购物体验。所以，可以说，这里的环境是建筑空间与自然空间的完美结合。设计师认为这是表参道大街独树一帜的特点，并在设计中延续了这一特点。

设计师希望在榉树下的街道空间与建筑上方的空间之间建立一种联系。他们在建筑上方打造了一个空中广场，让这座建筑与榉树融为一体。空中广场不仅让鸟儿与蝴蝶更容易飞到树冠上，而且整个空中广场已经成为一座屋顶花园，强烈地吸引着人们走进这栋建筑。众多购物者出现在楼顶的小广场上，也凸显了这一街区的商业氛围。

设计师首先对楼顶空间进行布局，设置了一系列平台，并在屋顶中央和四周栽种了树木。四周光线最好，并且能眺望十字路口的风景。然后，他们又打造了一个3层多高的外围结构。树木伸展出来的枝干填充了这一结构的内部空间。他们将屋顶上的平台架高，与种植树木的土壤层等高。就这样，设计师将不利条件进行巧妙利用，营造出亲切温馨的环境氛围，拉近了人与树的距离。平台上采用多边形台阶，造型类似于生长在树下的多孔菌类，这样就

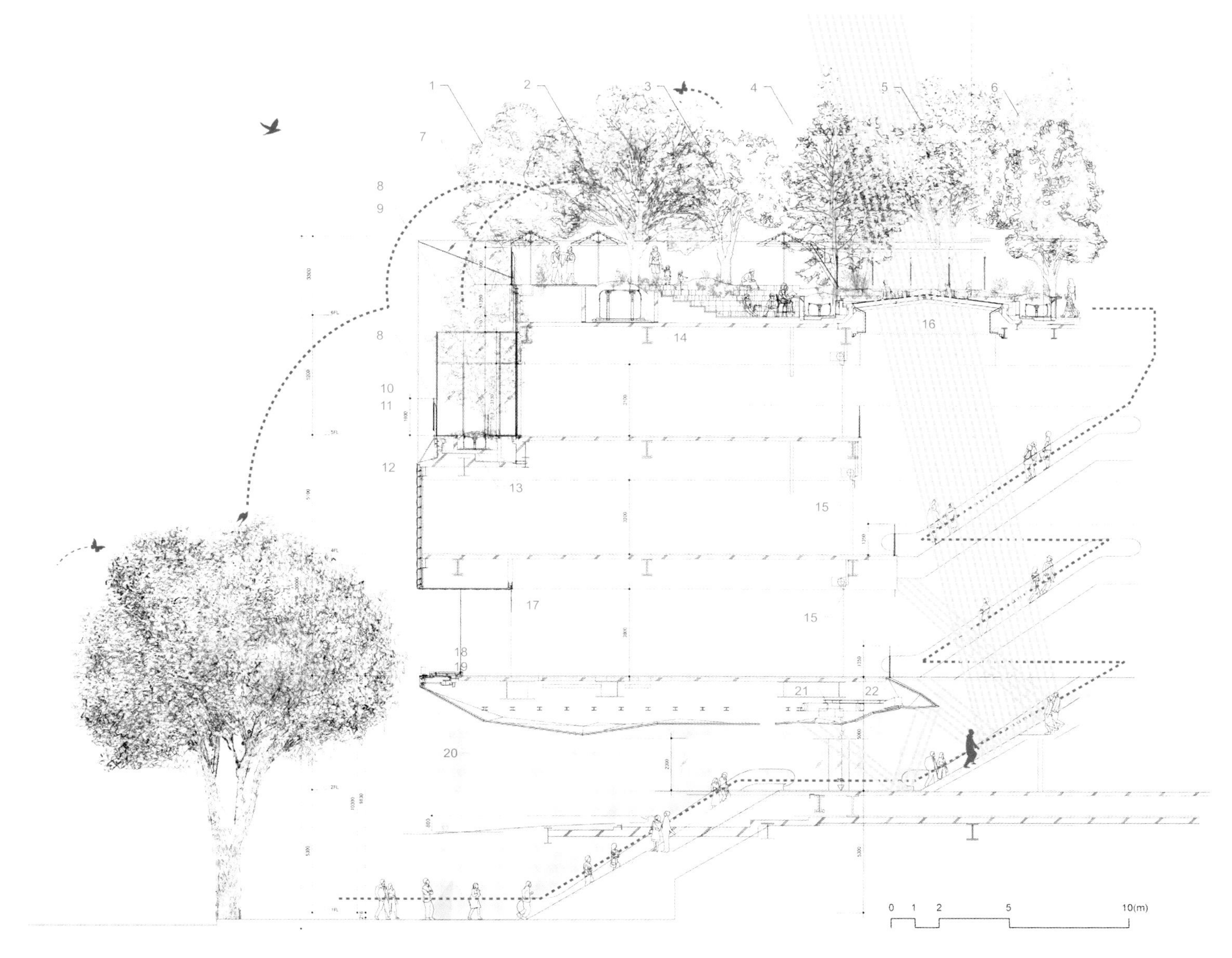

ELEVATION

1. Shirakashi
2. Camphor tree
3. Olive
4. Katsura
5. Japanese maple
6. Longstalk holly
7. Composite decking floating floor (structural steel)
   Protecting concrete t=80
   Heat insulating materials rigid urethane foam t=20
   Asphalt waterproofing
   External wall insulation glue method
8. Glass handrail post: st-FB
9. Fraxinus griffithii
10. Creeper Vinca Minor
11. Aluminium panel 3.0t acrylic resin
    Baking finish Enmel DIC-524 mid. colour
12. Extruded cement panel t75
    Aluminium press panel 2.0t
    Baking finish fluororesin DIC-524/525 mid. colour
13. Protecting concrete t60 protecting screed
    Asphalt waterproofing
14. Plant area:
    Composite decking floating floor
    Artificial light-weight soil
    Seepage sheet
    Drainage layer t=45
    Protecting concrete t=80
    Heat insulating materials rigid urethane foam t=20
    Asphalt waterproofing
    External wall insulation glue method
15. Smoke barrier automated rolling type
16. Skylight
17. Soffit: Aluminium press panel 2.0t
    Baking finish fluororesin DIC-524/524 mid. colour
18. Facility sign
19. Roof panel AL 2.0t
    Baking finish fluroresin
20. SUS 3.0t mirror finish
21. Shutter
22. Air curtain

立面图

1. 小叶青冈（日本常绿橡树）
2. 樟脑树
3. 橄榄树
4. 连香木
5. 日本枫树（鸡爪枫）
6. 冬青树
7. 合成材料浮式地板（钢结构）
   混凝土保护层（厚度：80）
   隔热材料（聚氨酯硬泡沫，厚度：20）
   沥青防水层
   外墙保温层
8. 玻璃扶手（st–FB）
9. 光蜡树
10. 藤蔓植物（小叶长春花）
11. 铝板（聚丙酸树脂，厚度：2.0）
    烘漆（DIC–524，中色）
12. 突出的水泥板（厚度：75）
    铝板（厚度：2.0）
    烘漆（氟树脂，DIC–524/525，中色）
13. 混凝土保护层（厚度：60，砂浆保护层）
14. 种植区：
    合成材料浮式地板
    轻质人造土壤
    渗漏层
    排水层（厚度：45）
    混凝土保护层（厚度：80）
    隔热材料（聚氨酯硬泡沫，厚度：20）
    沥青防水层
    外墙保温层
15. 隔烟层（自动伸缩型）
16. 天窗
17. 拱腹：铝板（厚度：2.0）
    烘漆（氟树脂，DIC–524/524，中色）
18. 指示牌
19. 屋顶铝制板材（厚度：2.0）
    烘漆（氟树脂）
20. 不锈钢（厚度：3.0，镜面抛光）
21. 遮板
22. 风帘

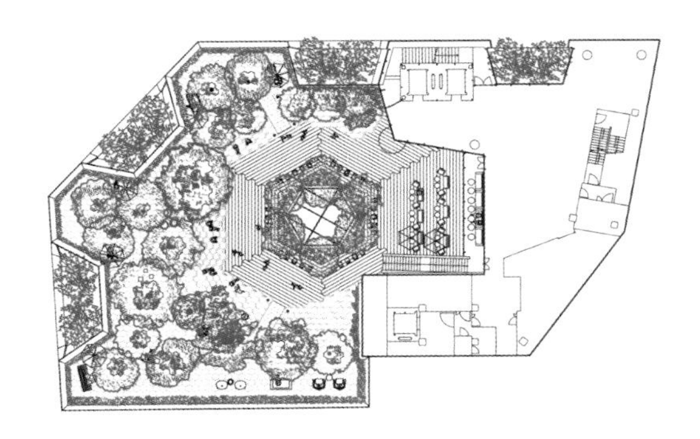

7F PLAN　7楼平面图

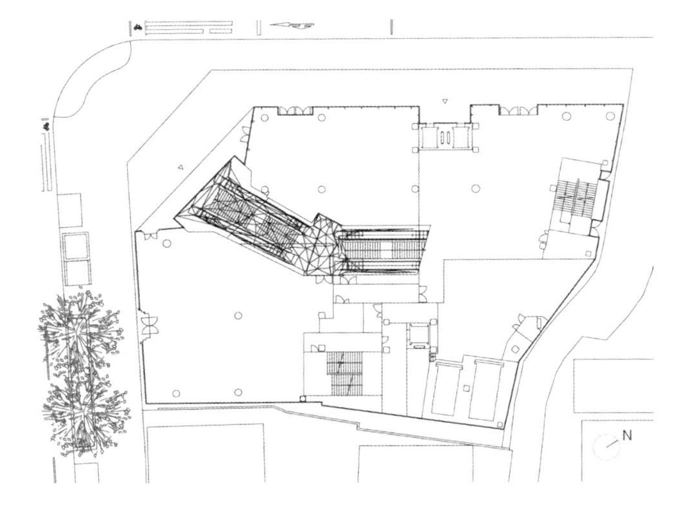

5F PLAN　5楼平面图

1F PLAN　1楼平面图

gives the wall a different appearance depending on how light strikes it. The functions demanded of the wall, in terms of performance and legal requirements, are borne by the extruded cement panels that form the backing. To this backing the aluminium polygonal panels were attached, using rotatable metal fittings designed to absorb seismic movement. The pressed-aluminium panels protrude in shape, with the peak of the protrusion off-centre. During its attachment, each panel was rotated 120 degrees to produce a composition of off-centre polygons having three different orientations.

Scenery reflected in the entrance tube, which opens on the intersection, changes with the season and time of day, and the tube presents a new appearance with each visit. To enable fine angle adjustment of the triangular mirror panels during their attachment, round pipes are employed as backing along the panel joints, and the panels attached to the pipes using hinges. Un

巧妙地解决了树木种植的土壤层造成的高差问题，同时造就了平台独特的造型。多边形台阶有多重功能，不仅是台阶，而且也可以作为桌椅来使用。人们通过这些妙趣横生的台阶与建筑进行亲密接触。

设计师发现，人们会自然地聚集到台阶下方，在广场中央形成一个中心。他们坐在不计其数的小台阶上，在小广场上坐成一圈，欣赏面前的美景。这种时候，你就会发现，无论现在网上购物有多么流行，现实中真实的购物体验还是不可取代的；现代商业空间的设计正是要去发掘其不可取代之处。在这样舒适的环境中享受轻松愉悦，伴随着独特的空间体验，这就是未来商业场所的设计最需要的东西。

万花筒式的入口设计

这栋建筑物的入口很特别，是筒状的，直达楼上。这一特色已经深深印在人们头脑中了，于是设计师保留了入口的基本造型。入口空间的镜面设计使其呈现出万花筒一般的炫目效果（镜面设计仿照中央公寓的镜面墙——中央公寓是从前地处这个位置的大楼，代表的是20世纪60年代和70年代的青年文化）。人们身着各色时装走在这里，映在无数镜中的景象真比万花筒还丰富，有种时尚界模特走T台的感觉。在入口乘自动扶梯变成了一次丰富多彩的体验。从门前的十字路口经过的路人都会忍不住向门内张望。随着扶梯上行，宽敞的中庭映入眼帘。阳光透过天花板上的天窗照射进来，由于屋顶上天窗周围种植了树木，摇曳的树影给中庭带来妙趣横生的光影效果。从这里，你将开始一段别样的购物之旅——完全符合表参道大街的特色。

利用光线打造千变万化的外墙

广场外围结构的外墙采用多边形组成的饰面材料，会根据阳光照射的角度不同而呈现出不同的外观。由于结构和法规方面的要求，外墙背面采用水泥板，用可旋转五金配件将多边形铝板贴在水泥板上（这种配件具有抗震功能）。每一小块多边形铝板都有一个偏

avoidably, the mirror panels will appear to leap out at one, but this assembly method enabled the designers to free them from geometric rigidity and form a spatial device that softly encloses people passing within.

A volume that merges greenery with architecture floats in the sky above Omotesando. In order to meld the building with the bustling shop-lined avenue, the ground floor tenant shops have each been allowed to freely design its own façade. As a result, each shop is distinct from the building's upper floors and neighbouring shops, and an impression of independent shops, each on its own site, is produced. Thus, Hiroshi Nakamura & NAP have fused commercial logic and the logic of environmental greenery in the design of the building.

离中心的凸起。施工时，每块铝板都旋转120度，拼合成由多边形构成的墙面，铝板的凸起朝向三个不同的方向。

映照在入口镜面上的景象随着季节和一天之中不同的时间而变化，所以你每次来到这里都会看到不一样的景象。为了在施工中保证三角形镜面的精确角度，设计师采用圆管来支撑镜面接头处，用折页来连接镜面与圆管。无数镜面的使用不可避免地会给人造成一种威胁感，但是设计师用这种安装手法，巧妙地打破了镜面刻板的几何造型，将入口变成一个柔和的空间元素，伴着人们走入大楼。

空中广场的设计将景观与建筑巧妙结合，成为悬于表参道大街上的一座空中花园。为了让这栋建筑融入繁华的商业街环境中，一楼的商铺可以自由地设计自己的店面。于是，每家店铺的门脸都跟建筑上层和附近的其他店铺不同，给人一种自由自在、千变万化的感觉。因此，设计师在这栋建筑的设计中将商业空间与环境绿化融为一体，实现了商业与绿化的糅合。

Trump Towers

川普大厦屋顶花园

Completion date:
2011
Location:
Istanbul, Turkey
Landscape Project:
DS Architecture – Landscape
Landscape architect:
Deniz Aslan, Günseli Döllük
Architect:
Brigitte Weber Architects
Photographer:
Gürkan Akay, Cemal Emden, Günseli Döllük
Area:
13,300 sqm

竣工时间：
2011年
项目地点：
土耳其，伊斯坦布尔
景观公司：
DS建筑事务所
景观建筑师：
德尼兹·阿斯兰，格文瑟里·德鲁克
建筑设计：
布里奇特·韦伯建筑事务所
摄影师：
戈尔坎·阿凯、杰马尔·埃姆登、冈扎利·杜鲁克
面积：
13,300平方米

Project description:

Trump Towers, one of the most significant investments located at the heart of business and urban life of Istanbul, is a unique project that has a smooth connection with the surroundings and is a natural part of the urban landscape through its car park and its connection to the subway.

Trump Towers are located in Şişli, a central location, in short distance to the bridge, and important districts such as Etiler, Levent and Beşiktaş. It rests on a total area of approximately 23,370sqm, 13,300sqm of it designed as open space.

The design which shapes both the urban and the private use-oriented outdoors and contains the conceptual design of the terrace and the courtyard of the towers, was based on a linear sequential setup that defined the utilisation of all the gardens.

The differences between the vegetational and rigid material, which were designed in a linear pattern, both facilitated a 3-D perception of the garden and provided a natural path within the garden. A horizontal mobility was obtained through the coupling of different patterns, and on the vertical, through the trees located by means of the elevated topography, through the linear walls and the covering materials. The seating elements, shaders, and the water elements brought diversity in the linear setup. The freshing and elevated feeling created by the two towers was repeated by means of the strong horizontality on the garden planes. The concept of the entire project aimed to simplify the perception of the spaces through a common design perspective.

While designing the entrance square that forms an interface with the urban life, it was planned to create an urban area that attracts

TRUMP
TOWERS
TRUMP TOWERS MALL
TOWERS
MALL
SATURN
MIGROS

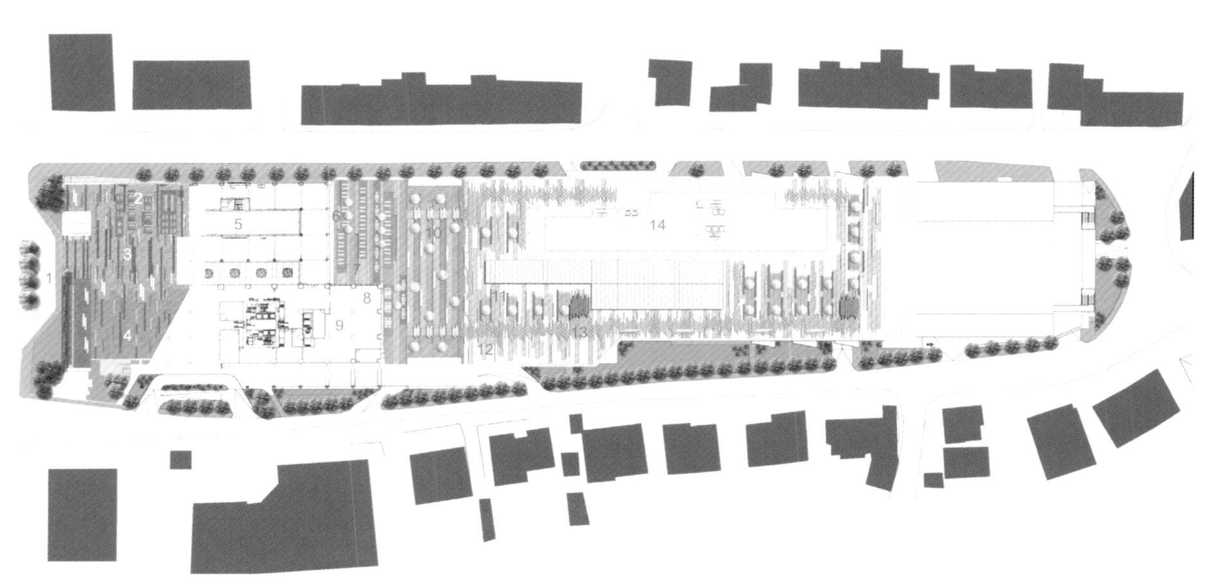

PLAN

1. Drop-off point
2. Café
3. Entrance square
4. Reflection pool
5. Indoor swimming pool
6. Sunbathing terrace
7. Elevation + 9.50 terrace
8. Café
9. Tower 1
10. Terraced garden
11. Elevation + 4.90 terrace
12. Mid-gardens
13. Walkway
14. Tower 2

平面图

1. 下车地点
2. 咖啡厅
3. 入口广场
4. 倒影池
5. 室内泳池
6. 日光平台
7. 平台（标高+ 9.50米）
8. 咖啡厅
9. 1号塔楼
10. 台地花园
11. 平台（标高+4.90米）
12. 中央花园
13. 走道
14. 2号塔楼

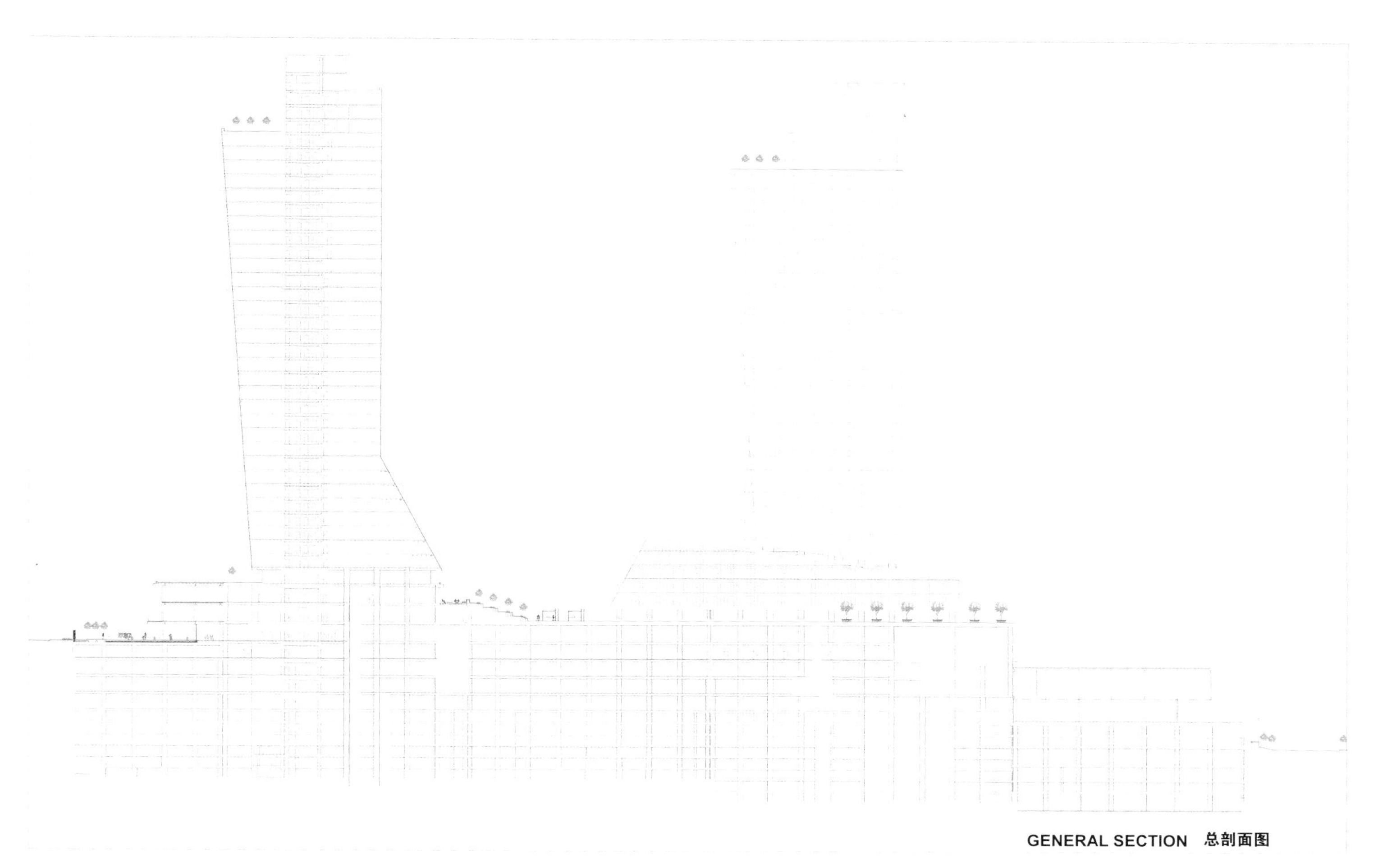

GENERAL SECTION 总剖面图

people to the building, and also that is used as a transition and a service area. The square became a meeting point with its catering platforms, its pool that matches the refraction of the ground, its green wall, which is also a symbol of eco-technology that filters out the noise and dirt of outside.

An impressive garden was aimed for in the terraces at the high levels that is far away from the city's chaos. It was intended that the linear lines, on which the design is based, is to be felt most intensively in this garden, where several utilisation possibilities were offered to the user as a multi-purpose open space. While one of the terraces was designed as a transition, walking and recreation area with the highest botanic intensity, the other terrace was designed as a garden carrying out the function of an open space and sunbathing area. The linear water elements in the sunbathing area were included in the design in order to cool the area in the hot weather and to create several reflecting surfaces.

The gardens of the inner spaces of Tower 1 and Tower 2, containing offices and residential areas respectively, are created without irrigation necessities. It was aimed to form a sustainable inner garden and included a small joke in the design. As planting material, imitations with striking colours were preferred and an abstract garden image was created.

While designing the lighting, it was planned to keep the linear impact both on the rigid floor and in the water by means of the light. Tall lighting elements were not preferred, instead the light that is reflected from the trees and the reflecting surfaces was used.

It was aimed that all the interwining functional spaces, which were designed according to a certain transition frequency; floor patterns, water elements, green components, and the lighting equipments generate different perspectives, all the while sustaining the character of a garden offering variety.

PLANTER DETAIL

1. a. Crushed stone no: 2-3 h: 6cm
   b. Geotextile 150gr./m^2
   c. Sand-gravel mixture 0-6.5cm
   d. Geotextile 150gr./m^2
   e. Drainage layer 25mm
   f. Protective geotextile and vapour barrier 500gr/m^2
   g. Water isolation (poliisobutilin membrane)
   h. Levelling concrete 4-12cm
   i. Thermal insulation xps 6cm
   j. Vapour barrier
   k. Reinforced concrete 10cm
   l. Steel carrier profile 400mm
2. a. Planting medium h: 6cm
   b. Geotextile 150gr./m^2
   c. Pumice stone 5cm
   d. Galvanised metal planter 4mm
   e. Geotextile 150gr./m^2
   f. Drainage layer 25mm
   g. Protective geotextile and vapour barrier 500gr/m^2
   h. Water isolation (poliisobutilin membrane)
   i. Levelling concrete 4-12cm
   j. Thermal insulation xps 6cm
   k. Vapour barrier
   l. Reinforced concrete 10cm
   m. Steel carrier profile 400mm
3. Uplight
4. Galvanised metal planter 150x150cm h: 67.5cm
5. Galvanised metal plate 4mm
6. Galvanised metal gusset plate 4mm
7. Water drain

种植槽细部图

1. a. 碎石（高度：6厘米）
   b. 土工织物（150gr./m^2）
   c. 沙子、碎石混合物（0～6.5厘米）
   d. 土工织物（150gr./m^2）
   e. 土工织物保护层和水蒸气
   f. 隔离层（500gr/m^2）
   g. 防水层（丁基橡胶膜）
   h. 混凝土找平层（4厘米～12厘米）
   i. 隔热层（挤塑板，6厘米）
   j. 水蒸气隔离层
   k. 钢筋混凝土（10厘米）
   l. 钢质载体（400毫米）
2. a. 栽种介质（高度：6厘米）
   b. 土工织物（150gr./m^2）
   c. 浮石（5厘米）
   d. 电镀金属种植槽（4毫米）
   e. 土工织物（150gr./m^2）
   f. 排水层（25毫米）
   g. 土工织物保护层和水蒸气隔离层（500gr/m^2）
   h. 防水层（丁基橡胶膜）
   i. 混凝土找平层（4厘米～12厘米）
   j. 隔热层（挤塑板，6厘米）
   k. 水蒸气隔离层
   l. 钢筋混凝土（10厘米）
   m. 钢质载体（400毫米）
3. 照明（向上照射）
4. 电镀金属种植槽（150厘米x150厘米，高度：67.5厘米）
5. 电镀金属板（4毫米）
6. 电镀金属角撑板（4毫米）
7. 排水管

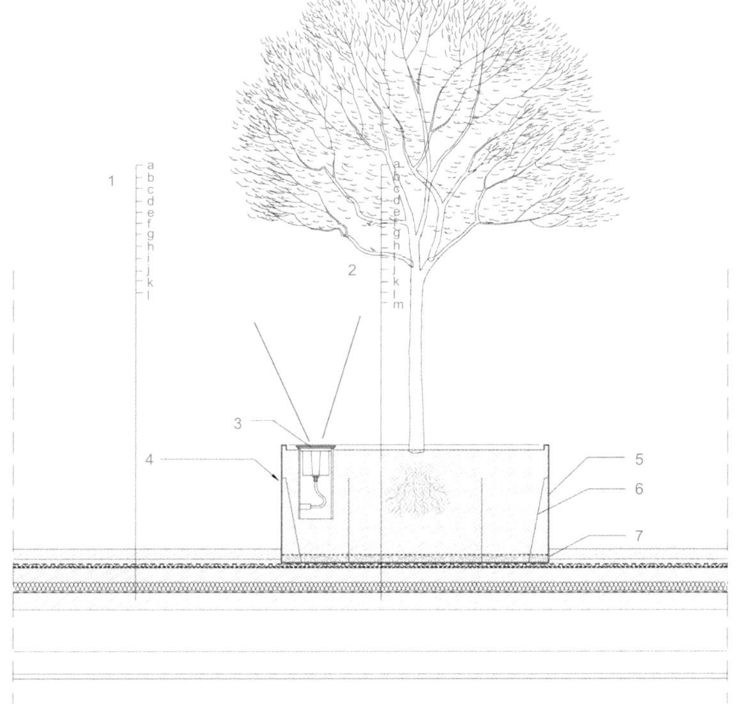

川普大厦位于伊斯坦布尔市中心最繁华的商业街区。大厦与周围的商业环境融为一体，楼下有停车场，与地铁相连，是伊斯坦布尔城市形象的一部分。

川普大厦所在的位置是希什利区，这是伊斯坦布尔最重要的中心区之一，离艾提雷区、莱文特区和贝西克塔斯区等几个重要街区都不远。川普大厦占地约23,370平方米，其中13,300平方米的面积设计成开放式空间。

开放式空间位于裙楼的屋顶，也是两栋塔楼的庭院。户外空间的设计既要满足公共空间的功能，又要满足个人使用的私密性。空间上采用线性布局，规划出一系列小花园。

屋顶设计采用植被和硬景观相结合，呈现出线性形态，二者形成鲜明对照，在三维效果的花园中形成一条自然的石材小路。三维花园包括水平和垂直两个方向。水平方向上，花园由多种几何形态拼合而成；垂直方向上，架高种植树木的位置，还有流线型的隔墙。坐具、遮阳棚和水元素的设计丰富了空间的线性布局。两座高高的塔楼有一种冲上云霄的气势，与下方水平方向上的花园相辅相成。屋顶花园的设计目标就是通过景观设计来改变人们对建筑空间的感知。

入口广场是川普大厦与城市环境的交汇点，这是一个过渡空间，其设计旨在把人们吸引到这里来。广场上有餐饮区，有泳池（水面上倒映着周围的景观），有绿墙（生态技术的象征，具有过滤外界噪声和污染的功能），是人们社交聚会的好去处。

屋顶花园的设计旨在打造一个远离城市喧嚣的静谧环境，一座优美的空中花园。花园的设计以几何线条为基础，划分出不同的空间，满足不同的使用需求。其中一个空间设计成散步和休闲空间，也是一个过渡空间，植被非常茂盛。另外一个空间采用开放式布局，侧重营造花园的氛围，适合日光浴。日光浴区的线性水景元素不仅通过水面倒映的景象丰富了空间的内容，而且能够在炎热的天气里降低花园的气温。

1号和2号塔楼内分别是写字间和公寓，室内也有绿化，并且无需灌溉，打造出可持续性的室内花园。设计上采用大胆的色彩，营造出抽象的花园形象。

灯光设计方面，设计师希望灯光能够突出花园地面和水景元素的线性布局。没有采用较高的照明设施，而是将照明灯安装在树上和水边。

花园内各个空间根据功能安排次序，彼此巧妙地衔接、过渡。地面铺装、水景元素、绿色植被以及照明设施带来多样化的感官体验，共同构成一座独特的空中花园。

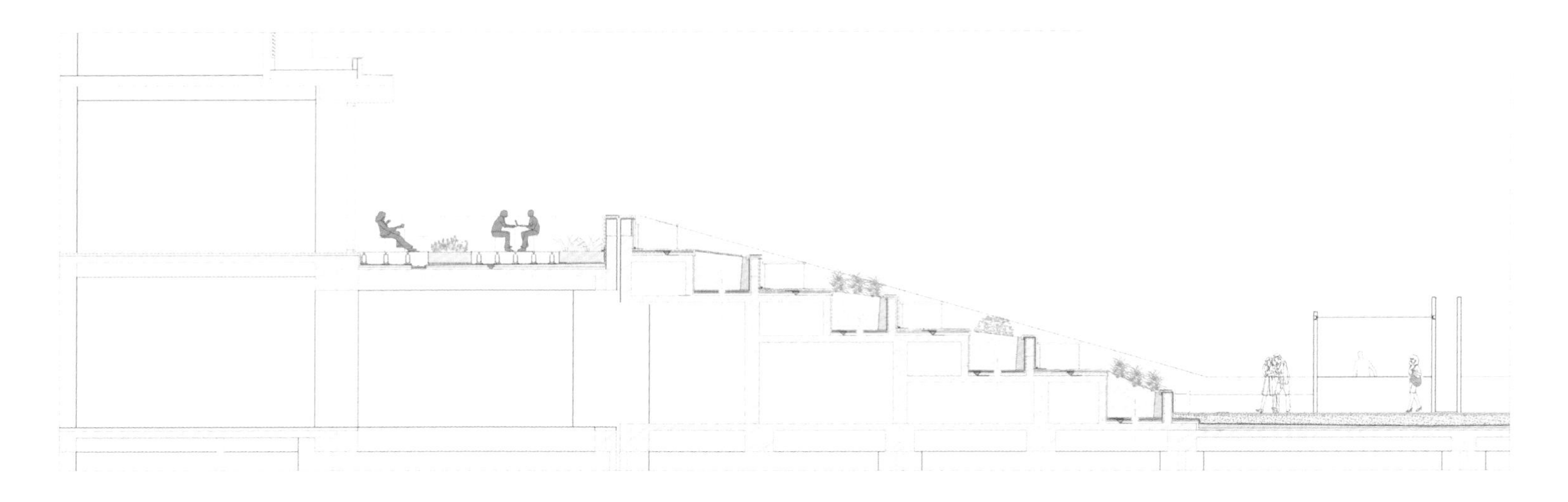

SECTION 剖面图

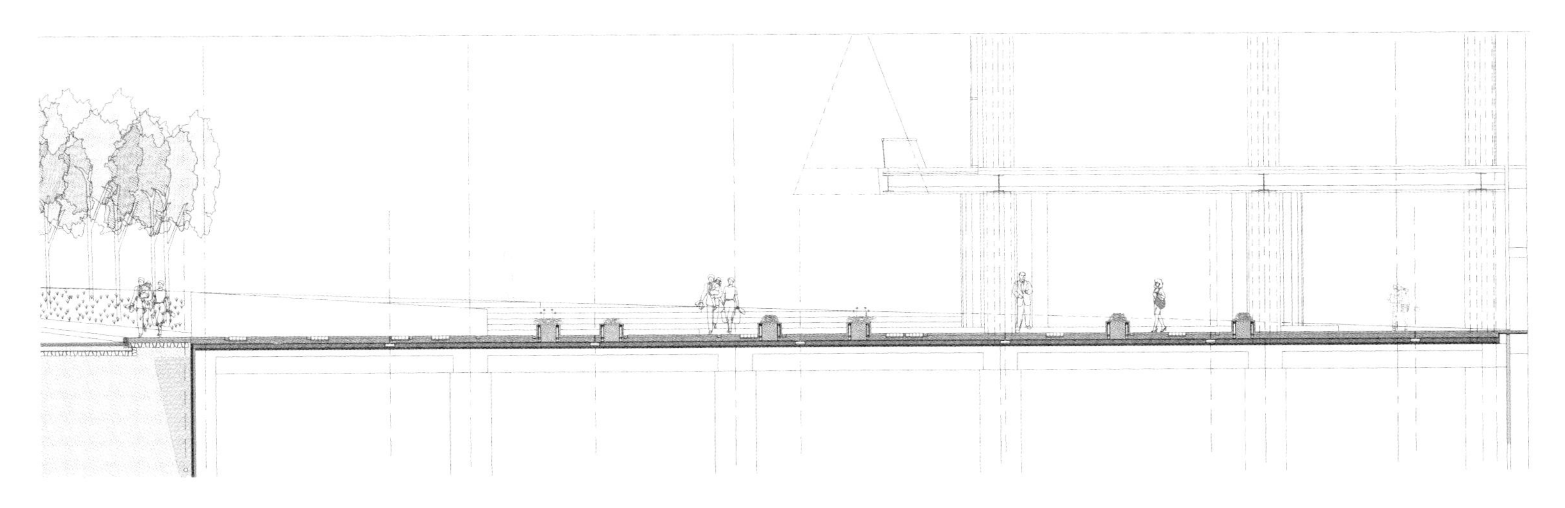

SECTION 剖面图

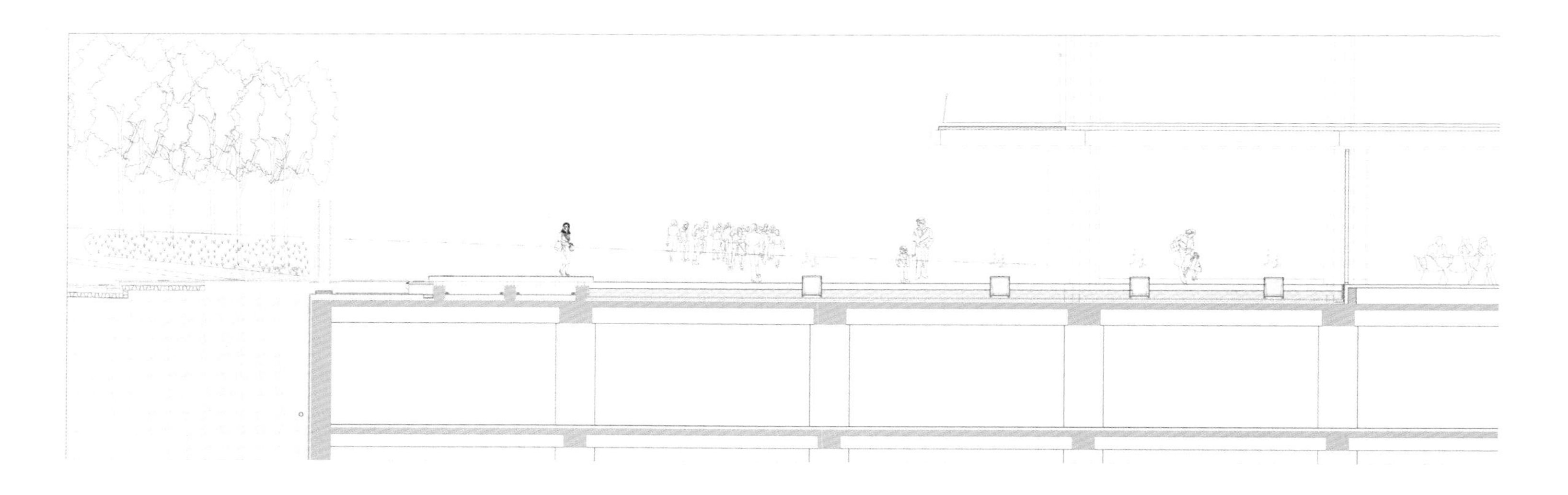

SECTION 剖面图

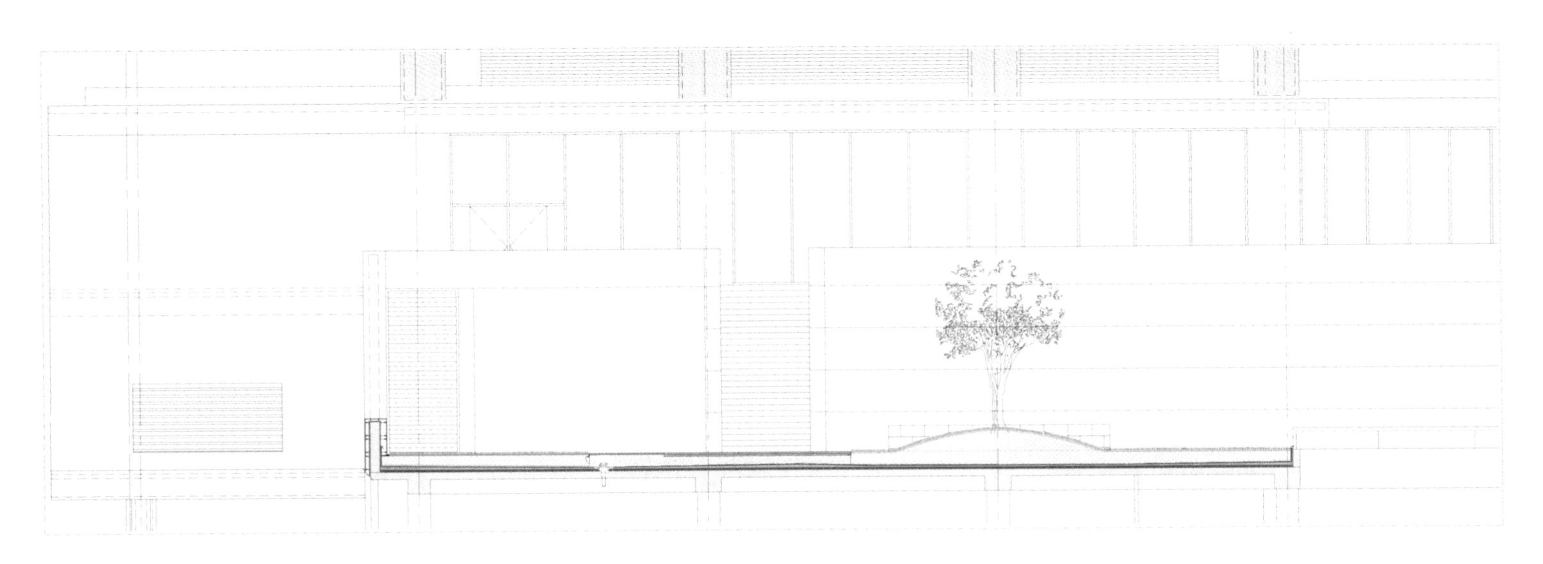

SECTION 剖面图

Westminster Terrace

香港皇璧公寓

Location:
Hong Kong, China
Architect:
Paul Davis and Partners
Landscape architect:
Andy Sturgeon Landscape and Garden Design
Client:
Grosvenor & Asia Standard
Photographer:
Andy Sturgeon
Area:
16,722 sqm

项目地点：
中国，香港
建筑设计：
保罗·戴维斯建筑事务所
景观设计：
安迪·斯特金景观园艺设计公司
客户：
高富诺集团、泛海国际集团有限公司
摄影师：
安迪·斯特金
面积：
16,722平方米

Project description:

Andy Sturgeon Landscape and Garden Design created the roof gardens and surrounding landscape at Westminster Terrace, a high-end development of 59 luxury apartments in Hong Kong.

The 33-storey tower, designed by Paul Davis + Partners Architects, sits above a cantilevered clubhouse and landscaped terraces with infinity pool and a tennis court. The pool deck was designed as a continuation of the interior design of the clubhouse and entrance foyers.

Andy Sturgeon's design creates a seamless connection between the indoor and outdoor spaces. Large vertical louvres extend into the landscape linking the gardens to the striking architecture, creating a strong visual backdrop to this generous outdoor space. The detailing is crisp, modern and elegant with strong influences from the luxury beach resorts of South-east Asia.

Andy Sturgeon explains: "The design was conceived to maximise views across the city skyline. Each landscape element was designed to respond to the strong architectural style of the building and create a dialogue with the internal spaces and materials."

Shaded pavilions for massaging or relaxing punctuate the garden to provide a variety of private spaces hidden away from the tennis court, play areas and entertaining zones. Throughout the garden, a range of materials and finishes such as dark and mid-grey granite, ceramic tiles and alpolic aluminium, provide contrasting colours and texture.

The planting scheme was selected to provide

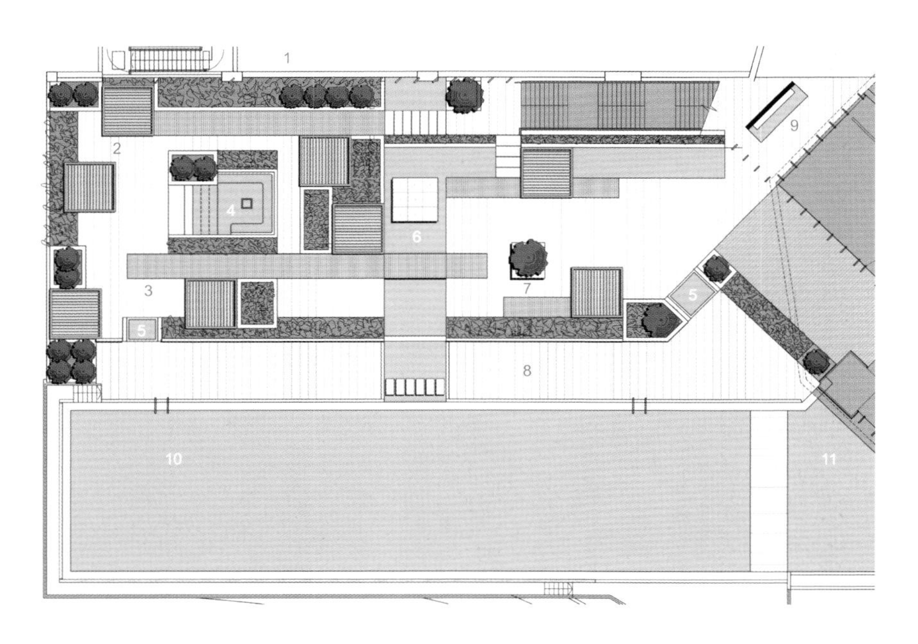

PLAN

1. Louvre panels
2. Pergola
3. Decking
4. Jacuzzi
5. Foot bath
6. Water feature
7. Planter
8. Sunbathing terrace
9. Water wall
10. Swimming pool
11. Children pool

平面图

1. 遮阳板
2. 藤架
3. 平台
4. 极可意按摩浴缸
5. 足浴区
6. 水景
7. 种植槽
8. 日光浴平台
9. 水景墙
10. 泳池
11. 儿童泳池

additional architectural interest. The designers chose plants that would suit Hong Kong's subtropical climate. Trees include Bambusa multiplex, Blucida molineti and Plumeria rubra while a wide selection of shrubs such as Jasminum grandiflorum, Duranta erecta, Nandina Domestica and Callistemon viminalis give structure at a lower level and ensure colour and interest throughout the seasons.

Since completing the Westminster Terrace project in Hong Kong, Andy Sturgeon has partnered with designers Jim Fogarty, Stephen Caffyn and Ronnie Tan to form Garden Design Asia, a new international garden design service created to offer the design styles and influences of British, Australian and Asian garden design for the growing Asian market.

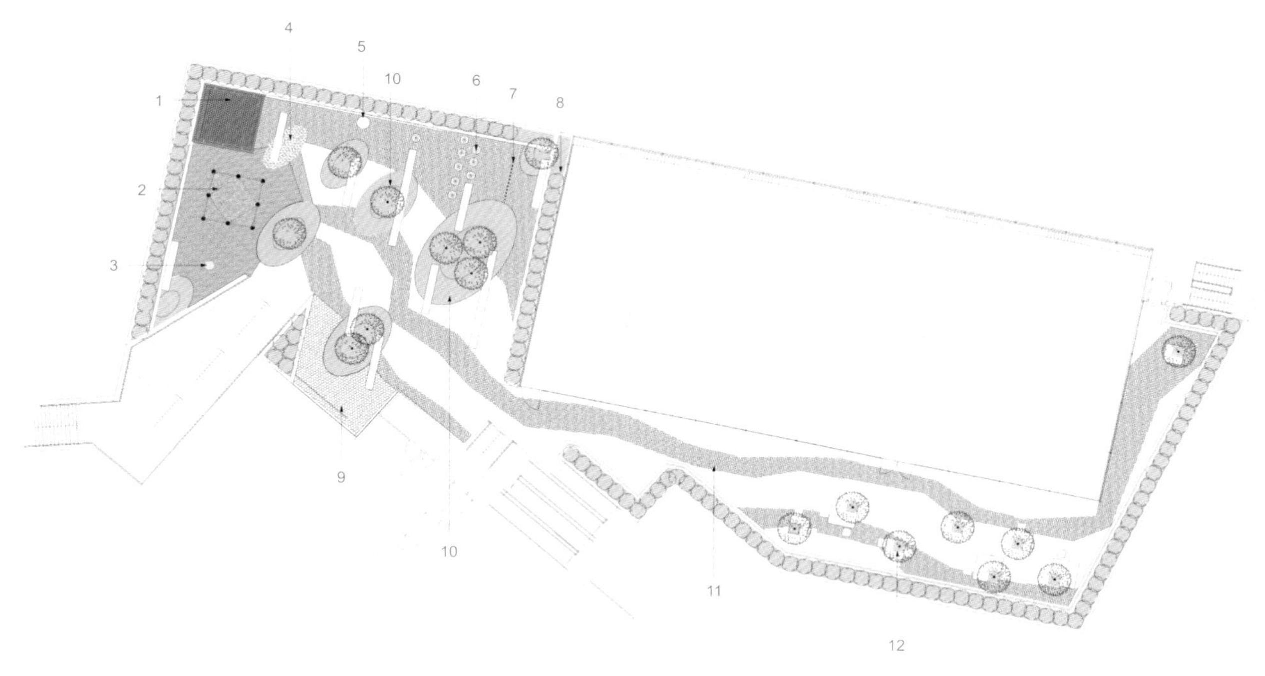

LAYOUT

1. Timber and steel shade pavilion
2. Geodesic spacenet grey supports with natural coloured rope
3. "Wobble dish" spinning play / Dutch disk
4. Dark play sand surface depth 400mm
5. Water spout with foot trigger
6. "Geyser" fountain jets
7. Spray jet bar
8. Evergreen screen hedging
9. Foot massage path
10. Planted mounds with long stacked stone benches cutting into earth form
11. Paving pattern picked out in contrasting paving materials
12. BBQ area with tree planters and built-in benches, plus tables and freestanding cube seating

平面布局

1. 木质和钢质遮阳棚
2. 灰色网格状承重结构（配有天然色的绳子）
3. “摇摆盘”
4. 游戏沙地（深度：400毫米）
5. 喷水嘴（配有脚动触发器）
6. 喷泉喷水口
7. 喷水槽
8. 常绿植物构成的树篱屏障
9. 足底按摩小径
10. 种植区（配有石质长椅）
11. 变换材质的铺装
12. 户外烧烤区（配有树木种植槽、嵌入式长椅、桌子和独立式立方体座椅）

香港皇璧高档公寓包含59个标准复式单元，屋顶花园以及周围景观的设计由英国安迪·斯特金景观园艺设计公司操刀。

这是一栋33层高的公寓大厦，建筑设计由英国保罗·戴维斯建筑事务所负责。公寓下方是俱乐部会所和景观台地，内有泳池和网球场。泳池边平台的设计是俱乐部及其入口门厅的室内设计的延续。

安迪·斯特金的设计将室内外空间完美地衔接起来。巨大的纵向遮光板一直延续到景观空间中，将建筑与景观融为一体，为宽敞的户外空间提供了强大的视觉背景。细节设计高雅而现代，没有一丝拖泥带水，体现出东南亚奢华的海滩度假建筑的风格。

景观设计师安迪·斯特金解释说："我们的设计旨在创造眺望香港天际线风景的最佳视野。每一个景观元素都与这栋公寓楼突出的建筑风格紧密相关，注重建立景观与室内空间和材料的关系。"

花园中设置若干遮阳亭，可以在这里做按摩或者休息。这样，除了网球场、游乐区和娱乐空间之外，公寓还提供了一系列私人休闲空间。统一的材料贯穿整个景观设计，如深灰色或者中灰色花岗岩、瓷砖和铝材等，带来色彩和材质的鲜明对比。

植被设计旨在为建筑添彩。设计师选择的都是能适应香港亚热带气候的植物。树木包括凤尾竹、红花缅栀等。低矮的灌木种类繁多，包括素馨花、金露花、南天竹、串钱柳等，修剪成各种造型，一年四季都呈现出缤纷的色彩和繁茂的景象。

做完香港皇璧这个工程之后，安迪·斯特金与吉姆·福格蒂、斯蒂芬·卡芬和罗尼·坦几位设计师一起创办了一家新的国际景观事务所——"亚洲园艺设计公司"，为不断增长的亚洲市场提供英国、澳洲以及亚洲园艺风格的景观设计。

Winery Terlan Wine Garden

泰尔拉诺酿酒厂屋顶花园

Location:
Terlan, Italy
Landscape architect:
el:ch landschaftsarchitekten
Client:
Kellerei Terlan
Photographer:
Christian Henke, Elisabeth Lesche
Area:
300 sqm

项目地点：
意大利，泰尔拉诺
景观设计：
el:ch景观事务所
客户：
泰尔拉诺酿酒厂
摄影师：
克里斯蒂安・亨克、伊丽莎白・莱施
面积：
300平方米

Project description:

In Terlano, the wine-making tradition dates back more than 2000 years. In the midst of this South Tyrolean wine region, the Cantina di Termal (Winery Terlan) was founded in the year 1893. With the extension of the traditional Cantina di Termal, one of the oldest wine cellars of South Tyrol, a roof garden in 300 sqm with a scenic view was added on top of the new outbuilding. The Germany landscape architects of el:ch landschaftsarchitekten designed this roof garden.

The wine garden is entirely dedicated to the experience of the spectacular landscape and thus closely connected to the locally produced wines. With its glass balustrades the garden merges seamlessly into the surrounding landscape.

A sequence of green carpets and wooden terraces inspired by local field structures is arranged on a surface covered in porphyry gravel. Together with the change in material one low step marks the border between moving and resting areas.

The plantation combines a low, carpet-like plain with stripes of perennial grasses. Seasonally emerging bulbs, such as Allium sphaerocephalon, add transient colour. Fragrant herbs and a pomegranate tree planted in a square metal container complete the sensory experience, linking the garden to agricultural tradition. Locally quarried porphyry, the warm colour of corten weathering steel and the lush green carpet combine to form a harmonious composition within a scenic setting.

Some plants were used for the roof garden, as well as
- Bulbs: Tulipa cretica, Allium sphaerocephalon
- Grasses: Sesleria autumnalis, Stipa tenuissima
- Perennial carpet: Sedum alba "Coral carpet", Chamaemelum nobile "Treneague"
- Solitary perennials: Achillea filipendulina "Hannelore Pahl", Eryngium x tripartitum, Helichrysum tianshanicum "Schwefellicht"

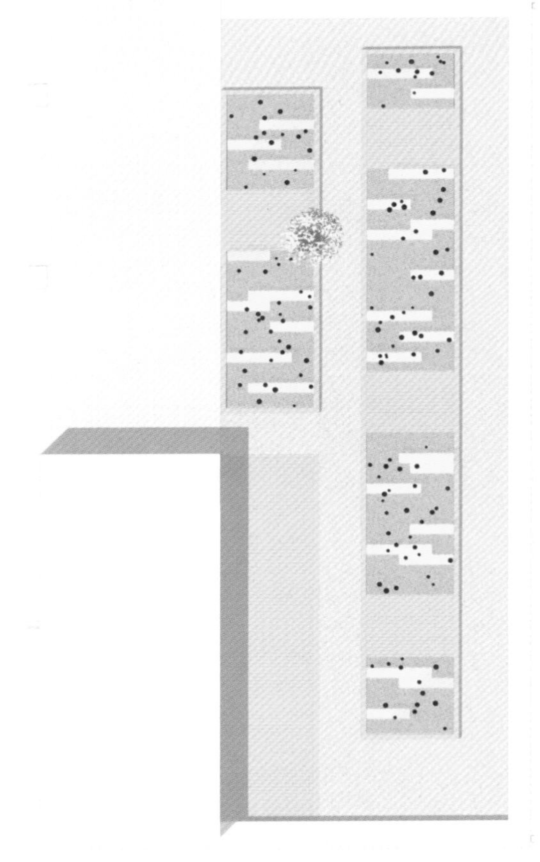

PAVED SURFACES　铺装地面

Gravel　砂砾

Wooden deck　木板平台

VEGETATION SURFACES　植被地面

Perennial ground covers　多年生地表植被

Ornamental grasses　观赏性草本植物

Ornamental bulbs　观赏性鳞茎类植物

Pomegranate tree　石榴树

SECTION 剖面图

SECTION **剖面图**

意大利泰尔拉诺的酿酒历史可以追溯到2000多年前。泰尔拉诺酿酒厂始建于1893年，是南蒂罗尔省最早的一批酿酒厂之一，周围风景秀美。2009年，酒厂进行了扩建，本案中300平方米的花园就建在扩建部分的屋顶上。德国el:ch景观事务所负责这座屋顶花园的景观设计。

屋顶花园的设计旨在突出酒厂周围的自然风景，这样也就与当地的酿酒文化建立了联系。设计师在屋顶四周设置了玻璃护栏，让屋顶花园自然地融入周围景观。

设计师受当地景观形态的启发，将绿毯般的草皮和木质平台穿插设置，空隙中铺设斑岩碎石。低矮的台阶划分出步行区和闲坐区的界限，不同空间采用的材质也相应发生变化。

屋顶花园的植被设计包括绿油油的草皮，其中种植了一行行的多年生草本植物。鳞茎类植物随季节生长，如圆头大花葱，为花园带来色彩的变化。正方形金属花坛里有一棵石榴树，还有气味芬芳的草本植物，进一步丰富了屋顶花园的感官体验，呼应了当地的农耕传统。斑岩是从当地采集的，耐候钢带来温暖的色彩，植物生长得非常茂盛。这些元素加起来，共同构成和谐一体的花园景致，与周围的自然风景相得益彰。

屋顶花园中采用的植物种类包括：
- 鳞茎类植物：郁金香、圆头大花葱
- 草本植物：秋兰草、细茎针茅
- 多年生草本植物：白景天（“珊瑚毯”）、罗马洋甘菊
- 单生多年生植物：凤尾蓍、三裂刺芹、蜡菊

Wyne Sukhumvit Condominium

素坤逸韦恩公寓

Completion date:
2012
Location:
Bangkok, Thailand
Architect:
DB Studio
Landscape architect:
Sanitas Studio Co., Ltd.
Client :
Sansiri PLC
Photographer:
Chaichoompol Vathakanon
Area:
4,293 sqm (5th floor: 713 sqm; 27th floor: 158 sqm)

竣工时间：
2012年
项目地点：
泰国，曼谷
建筑设计：
DB设计工作室
景观设计：
萨尼塔设计工作室
客户：
盛诗里公司
摄影师：
卡姆普・瓦萨卡南
面积：
4,293平方米（5层：713平方米；27层：158平方米）

Project description:

Background

In 2010, Sanitas Studio was commissioned by Sansiri PLC to do landscape design for a high rise condominium on Sukhumvit Road. Apart from landscape design for facility level and garden on ground floor, Sanitas Studio was asked for landscape design approach on the façade of the four-storey carpark building.

Located on Sukhumvit Road, the main road lining from city centre to the other part of the country, Wyne Sukhumvit situates at the transition of fast developing area. The fast developing area is a mixture of old shop houses, new developments and convenient transportation, but it is far from recreation areas or public parks in 1 kilometre radius. Living in a big city, sometimes creates stress and routine lifestyle.

Sanitas Studio's proposal was to consider the site as a Sanctuary in itself:

- A Sanctuary which is contrast from its surroundings as a jungle of concrete;
- A Sanctuary which provides the green, fresh air, morning drops;
- A Sanctuary in which residents can feel safe and secure;
- A Sanctuary which can brighten up residents' mind from the routine lifestyle;
- A Sanctuary which can bring the residents home;
- A Sanctuary which can bring other dimension to their life; Art, Culture and Nature, and create a garden pause from the city where people can rest, relax, contemplate and retreat from the city life.

Sanitas Studio aims to integrate planting in

Re - entry units
2.59
9 Units. Only
2.59
Re - entry units
Only 9 Units.
02-711-1133

PLAN 平面图

every possible area including a garden walk around the building foot print, a herb garden within a sculpture garden, vertical garden on façade of the building and rustic garden on swimming pool level. The idea was to encourage people to see the importance of nature and how it could improve their quality of life.

设计背景

2010年，盛诗里公司委托萨尼塔设计工作室为他们开发的一栋高层公寓楼做景观设计。这栋大楼位于曼谷素坤逸路上，除了设备层和一层的景观设计之外，还包括四层高的停车楼的外立面绿化。

素坤逸路是连接曼谷市中心与外界的一条重要公路。韦恩公寓位于飞速发展的市区边缘。这个街区内有古老的独栋店铺，也有新建的大楼，交通十分方便，但是离休闲娱乐区和公园较远（约1公里）。生活在大都市里有时会带来过大的压力以及日复一日单调的生活方式。

SECTION
1. Planting
2. Driveway
3. Parking
4. Sculpture lawn
5. To drain
6. Boundary line

剖面图
1. 植被
2. 车道
3. 停车场
4. 雕塑草坪
5. 排水
6. 边界线

Landscape Concept

Revealing

The landscape design was expressed in the form of Revealing of Sanctuary: the meeting point between nature and manmade. This design emphasises the importance of nature in urban life and questions the current balance between them.

萨尼塔设计工作室的设计旨在将这栋公寓打造成一处世外桃源：
包围在周围林立的混凝土建筑物之中的世外桃源；
有着绿色景观和新鲜空气的世外桃源；
让住户感觉安全舒适的世外桃源；
让住户跳脱出日复一日的单调生活的世外桃源；
让住户真正感到家一般温馨的世外桃源；
将艺术、文化与自然融入生活的世外桃源。
在这里，人们可以暂时远离紧张的城市生活，在花园般的环境中休闲放松。
萨尼塔设计工作室尽量在每个角落都加入绿色景观元素，包括：围绕着公寓大楼而设的散步道；雕塑花园内部又设置一个小花园；公寓大楼外立面上的“垂直花园”；泳池边的花园。整体设计思路就是让人们关注大自然，让人们看到大自然是如何改善我们的生活品质的。

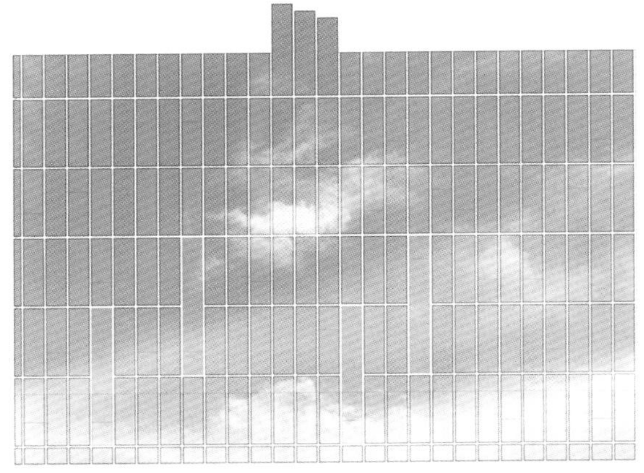

THE MODULAR FAÇADE STRIP 条状模块外立面

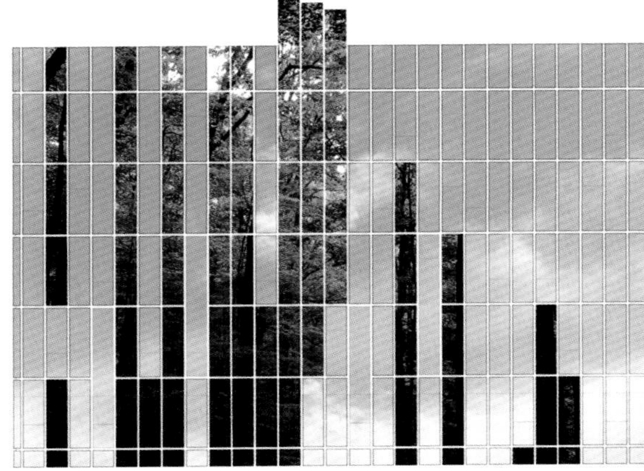

THE REVEALING 剥除建筑表皮，代之以绿色植被

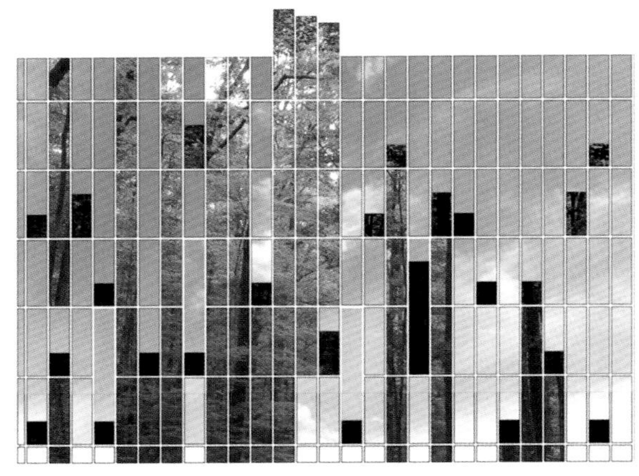

THE REFLECTION 植物倒映在外立面上

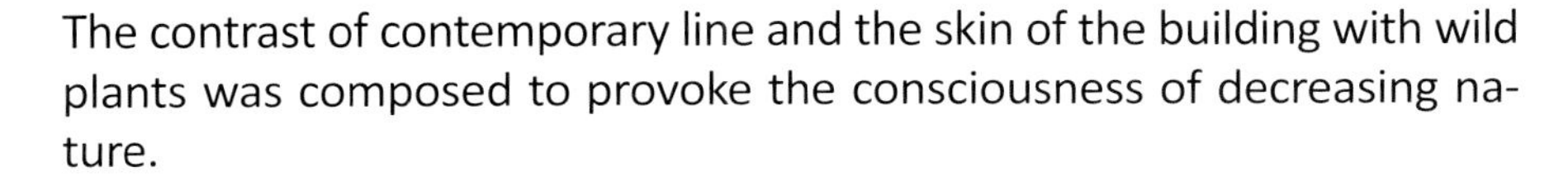

The contrast of contemporary line and the skin of the building with wild plants was composed to provoke the consciousness of decreasing nature.

"The Revealing" was in an action of peeling architectural skin. The façade was revealed by folding up stainless steel strips to reveal the vertical sanctuary. The pool was revealed by peeling strips of timber trellis to reveal the sanctuary. The sculpture garden was revealed by crafted topography.

Site Plan

At Wyne, Sanitas Studio has created a series of rich detailed garden spaces from the front plaza, a garden walk around the building, sculpture garden, swimming pool garden on 5th floor and sky garden on 27th floor.

At Wyne, soft planting also plays the important roles: nurturing the space, providing shade for outdoor function space and decreasing temperature, which is crucial in tropical country. Located in city centre, the space is limited and under strict regulations. There is 6 metres setback around the building for fire engine access. The area for actual garden is in the corner at the west end of the site. The challenge is to create the garden experience within limited space. Using turf pave on fire engine access is the idea to encourage the garden walk visually.

Sanitas Studio also proposed a vertical garden on the front façade of the four-storey carpark building in order to create a better environment for the surrounding neighbour and decrease the temperature of the building by buffering the glaring heat, and providing ventilation at the same time.

The landscape areas were located on three storeys: Gardens on Ground Floor, Swimming Pool garden on 5th Floor, and Sky garden on 27th Floor.

Gardens on Ground Floor

On ground floor, the whole plan was covered by both strips of softscape and hardscape. This created one large garden pattern that wrap around the building foot print.

景观设计理念

剥除人工，展现自然

景观设计通过层层剥除人工痕迹，呈现出一个天然的世外桃源，实现了天然与人工的完美结合。设计师强调大自然在城市生活中的重要性，对当今自然与城市生活二者之间的失衡予以关注。

设计师通过采用野生植物，突出了现代建筑的线条和表皮与自然的鲜明对比，让人们意识到正在逐渐萎缩的大自然。

剥除冰冷的建筑表皮，代之以绿色植被。

剥除外立面的不锈钢，代之以垂直花园。

剥除泳池边的木质框架，营造世外桃源。

剥除光秃秃的地表，打造绿色地貌的雕塑花园。

景观规划

韦恩公寓的景观设计包括一系列的花园空间：正门小广场、围绕着大楼的散步道、雕塑花园、5层的泳池花园以及27层的空中花园。

绿色软景观对韦恩公寓的环境起到重要作用：丰富了空间，为户外空间带来阴凉，降低温度（这在热带国家尤为重要）。由于地处市中心，面积有限，相关规定也十分严格。为了便于消防车出入，建筑四周留有6米的空地。真正的花园空间位于场地西端的角落里。设计面临的挑战是如何在有限的空间内营造花园般的空间体验。设计师在消防车通道上采用草皮，从视觉上营造出绿色散步道的感觉。

萨尼塔设计工作室还为四层高的停车楼设计了外立面的绿化，打造了一座“垂直花园”，不仅为附近居民营造了更加美好的环境，而且还起到降低建筑温度的作用，因为大楼表面的绿色植被能够缓冲直射的太阳光，同时对楼内通风也有帮助。

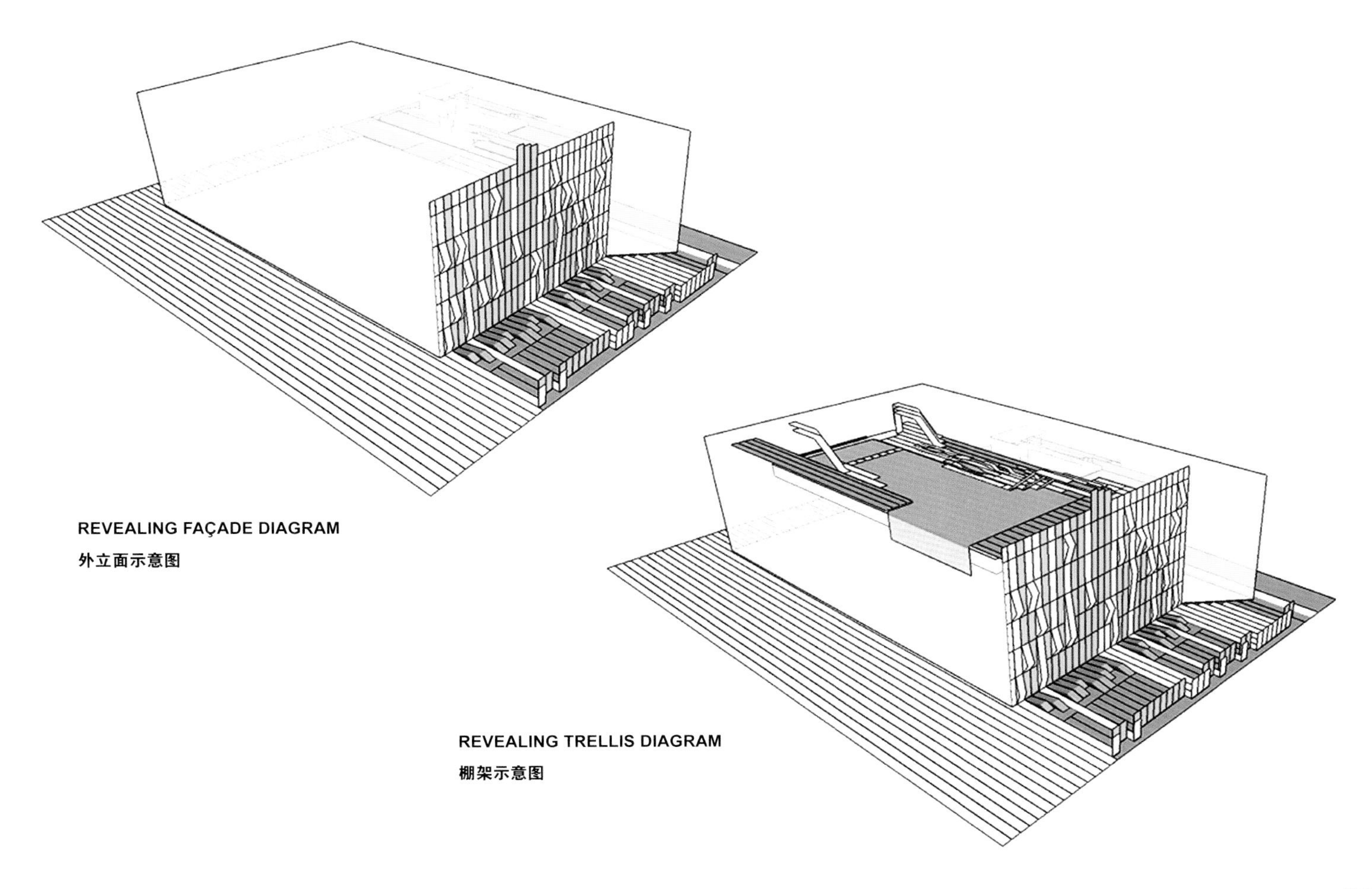

REVEALING FAÇADE DIAGRAM

外立面示意图

REVEALING TRELLIS DIAGRAM

棚架示意图

The sculpture garden is located in the corner, which provides the quietness and contemplation space for the residents. In the sculpture garden, the earth was pushed to reveal the water and its topography profile. The crafted berm was marked by stainless steel wall and showed its topography profile. The mirrored finish of the wall marked the existence of manmade structure in nature. At the same time, the reflection of nature on it made the manmade structure disappear.

The reflection, the meeting point between manmade and nature, is the key element in this sculpture garden. It penetrates in the garden, in the same way as man acquires space to live with nature.

景观设计分布在三个楼层上：一层花园、5层的泳池花园以及27层的空中花园。

一层花园

一层花园的布局采用硬景观和软景观相结合，共同构成整体的景观形态，将公寓大楼包围在花园环境中。

雕塑花园位于角落里，环境十分静谧。里面的地面经过整饰，营造出植被和水景结合的独特地貌。水景中采用不锈钢板，表面反光，突出了自然环境中的人工痕迹。同时，不锈钢板上映出周围的自然景色，又让这一人工雕琢的元素不着痕迹地消失了。

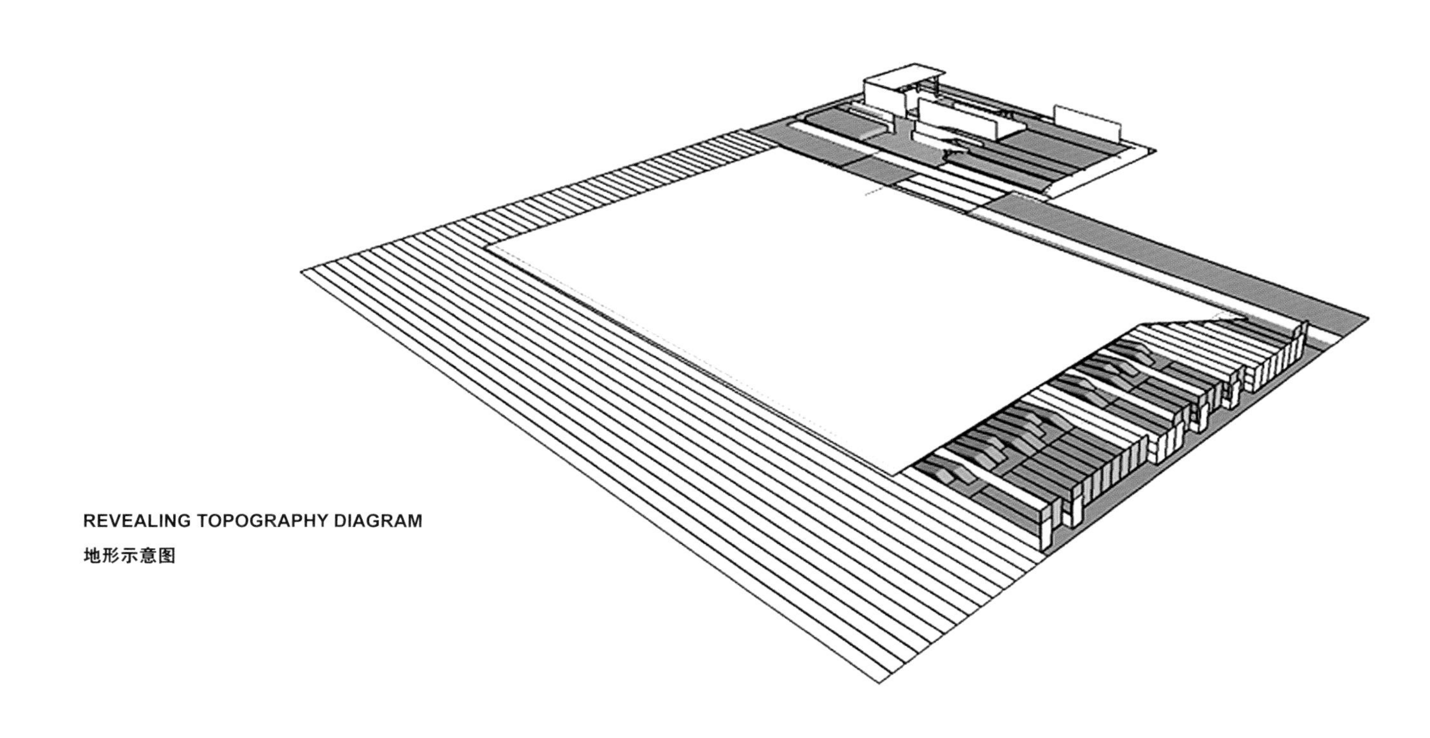

REVEALING TOPOGRAPHY DIAGRAM
地形示意图